DESIGNING NETWORKS
WITH CISCO

DESIGNING NETWORKS WITH CISCO

H. PASRICHA

D. JAGU

CHARLES RIVER MEDIA, INC.
Hingham, Massachusetts

Acquisitions Editor: James Walsh
Cover Design: The Printed Image

CHARLES RIVER MEDIA, INC.
10 Downer Avenue
Hingham, Massachusetts 02043
781-740-0400
781-740-8816 (FAX)
info@charlesriver.com
www.charlesriver.com

This book is printed on acid-free paper.

Harpreet Pasricha and Dattakiran Jagu. *Designing Networks for Cisco.*
ISBN: 1-58450-345-9

Library of Congress Cataloging-in-Publication Data

Pasricha, Harpreet.
   Designing networks with Cisco / Harpreet Pasricha and Dattakiran Jagu.
      p. cm.
   ISBN 1-58450-345-9 (pbk. with CD-ROM : alk. paper)
   1 Computer networks--Design and construction.   2. Cisco IOS.   3. Local area networks (Computer networks)   4. Wide area networks (Computer networks)   I. Jagu, Dattakiran. II. Title.
   TK5105.5.P469 2004
   004.6'5--dc22

                                                                    2004005013

Printed in the United States of America
04 7 6 5 4 3 2 First Edition

Contents

1 Technology Overview

When the dimensions of an organization increase, the need to connect multiple networks also increases. The collection of individual networks connected together using diverse technologies, devices, and protocols is known as *internetwork*. Designing internetworks with Cisco requires an in-depth knowledge of networking technologies. It also requires a practical approach, which comes from a clear understanding of the implementation of these technologies.

OSI REFERENCE MODEL

For many professionals and beginners in the field of networking and IT-enabled services, the Open System Interconnection (OSI) model has proved to be an industry standard, and a constant reference while working with any networking technology. Any networking technology is defined as a combination of networking devices, applications, and protocols. All these components can fit into a model that helps in providing a reference point for the industry.

A *model* is known as a reference point to which designing and planning of a technology can be done. The networking industry has to accept a model to develop components of technology so that maximum interoperability can be achieved. Interoperability can be possible only when everybody follows a set of standards or a model such as the OSI reference model.

We will discuss this model in detail, and understand the exact working of its architecture. This section is an important aspect for the day-to-day work of a networking professional.

The OSI reference model sets standards describing how data communication can be achieved between two endpoints, for example, two computers in an internetwork. To achieve successful transmission of data, this model divides networking technology into seven layers. Each layer performs a certain set of functions to accomplish this task. Layering reduces the complexity of data transmission. The software utilized at these layers includes operating systems, applications, and protocols, and hardware includes connectors and NICs. The seven layers in the OSI reference model are shown in Figure 1.1.

You can memorize the sequence of the layers of the OSI reference model using the first letters of the words of the sentence "**A**ll **P**eople **S**eem **T**o **N**eed **D**ata **P**rocessing,"

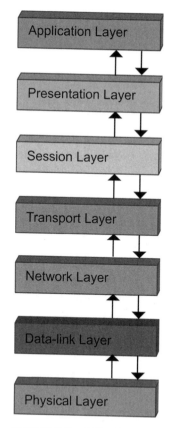

FIGURE 1.1 OSI reference model showing the seven layers.

that is, APSTDP for *Application, Presentation, Session, Transport, Data Link*, and *Physical*. The OSI reference model states that data flowing from source to destination must always pass through each layer of the model. Figure 1.2 displays the flow of data from the source to destination.

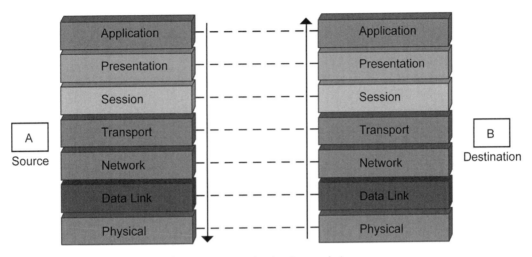

FIGURE 1.2   Data movement from source to destination and vice versa.

In Figure 1.2, consider "A" as the source and "B" as the destination. According to the OSI reference model, any form of data would start from the Application layer of the source "A" and travel down to the Physical layer. After reaching the Physical layer of the destination "B," the data moves down to the Application layer of "B." At each layer, the data or message is modified according to the functionality of each layer. As the data is processed, each layer adds a header and a trailer to its data (which consists of the next higher layer's header, trailer, and data as it moves through the layers). The headers contain information that specifically addresses layer-to-layer communication. For example, the Transport Header (TH) contains information that only the Transport layer sees. All other layers below the Transport layer pass the Transport Header as part of their data.

The benefits of a layered protocol model are:

- Interoperability of products from different vendors.
- Programming interfaces can be easily modified.
- The model is structured because each layer introduces headers and trailers as the data moves across the different layers.
- Each layer performs a specific function.
- Troubleshooting is simplified.
- Enhanced compatibility.

- Future upgrade of a single layer does not affect other layers.

Functions at Each Layer

The OSI reference model is a conceptual model composed of seven layers, each layer specifying particular network functions. The OSI model divides the tasks involved with moving information between networked computers into seven smaller, more manageable task groups. A task or group of tasks is then assigned to each of the seven OSI layers.

The next section details the functions of the seven layers of the model.

Physical Layer

The physical layer is considered to be Layer 1 of the OSI model. The prime function of this layer is to interact with the transmission media and to put the data on the media in the form of bits. The Physical layer of source "A" will transmit the bits, and the Physical layer of destination "B" will receive the bits. The main task for the Physical layer is to put the stream of bits on the media, irrespective of its type. All cables and devices such as repeaters and hubs reside on this layer. The Physical layer also deals with transmission through pins of a connector and physical interface of the NIC card. This layer also performs a function known as *signal encoding*. An important feature of this layer is bit formation. This means that the Physical layer considers the data in the form of bits.

It is often understood that the NIC card functions at Layer 1. This is not true since the NIC card contains the Media Access Control (MAC) address, which works at Layer 2 of the OSI. This will be discussed in the Layer 2 section.

The functions of the Physical layer are:

- Determines how electrical signals should be transferred across physical media
- Controls modulation, demodulation, and decoding of electrical signals
- Converts bits into electrical signals
- Deals with all physical aspects of a network

Data-Link Layer

This layer has two sub-layers:

- Media Access Control (MAC)
- Logical Link Control (LLC)

The Data-link layer considers the data in the form of frames, and its main function is to scramble and assemble the frames into bits for use by the Physical layer at the source and the destination. The MAC sub-layer provides an address for each device on the network while the LLC establishes and maintains links between the devices communicating with each other. The MAC addressing function provides a unique address to each device present in any network. Every NIC has a unique hardware address, known as a *MAC address* or *physical address*. This address is a unique number assigned to the NIC, and contains a hexadecimal number consisting of numbers 0–9 and letters a–f, for example: 12:36:ab:66:cd:88. You will learn more about MAC addressing in the section "LAN addressing."

Table 1.1 shows the different parts of the Ethernet frame and their functions.

TABLE 1.1   Parts of a Frame

Parts of a Frame	Functions
Preamble	• Known as the Start indicator • Marks the beginning of a data frame
Destination MAC Address	Stores the address of the destination node
Source MAC Address	Stores the address of the source node
Type	• Known as the length field • Indicates the type of data
Data	• Contains the actual message • Contains information about the above layers
Frame	• Indicates the status of the frame, such as any damage or corruption of data • Performs the Cyclic Redundancy Checksum (CRC) function

The functions of the Data-link layer are:

- Sends data across a particular link or medium
- Specifies physical addressing, sequencing of frames, and flow control
- Sets control methods to prevent data bottlenecks and traffic congestion
- Performs error detection and error correction
- Counts the frames to ensure that there is no loss of data packets
- Organizes packets in correct sequence order before transmission

Network Layer

This layer provides another type of addressing, known as *Network layer addressing* or *Layer 3 addressing*. This is also called "logical addressing." This type of addressing is protocol specific. The Network layer logically splits the network. This layer specifies a mechanism to deliver *packets* (also known as datagrams, packets are units of data at Layer 3) from one network to another, which forms an internetwork. This is done using the best possible path from source to destination.

No physical encoding takes place in Layer 3. Only logical addressing is required to set the path. This is done with the help of the routing table according to the configured network address. A *routing table* contains the reachability information to all destinations. Different routing protocols populate this routing table.

The functions of the Network layer are:

- Transmits data between devices that are not locally connected
- Controls congestion and internetwork traffic
- Performs encapsulation of data packets before transmission (covered later in this chapter)

Transport Layer

This layer provides reliability to the data delivery mechanism of an OSI reference model. The Transport layer divides the data into segments, which are easier to deliver and track. This layer uses "sequencing" to deliver segments to the destination. During sequencing, the Transport layer assigns numbers to each segment and asks for acknowledgement of each received segment from the destination. If the destination does not give an acknowledgement for a missed segment, this layer helps in retransmitting that particular segment. This is also called the *connection-oriented* feature of the Transport layer. The protocol for this feature is Transmission Control Protocol (TCP). In addition to TCP, the User Datagram Protocol (UDP) is a connectionless protocol and is responsible for transferring data packets between two interconnected computers in a network. It also provides port numbers, which help in recognizing the user requests.

The Transport layer provides error checking at the segment level, ensuring an error-free host-to-host connection. At the receiving end, the segments are checked for errors. If segments are received without errors, they are reassembled in the proper sequence and an acknowledgement is sent to the host. If the host does not receive an acknowledgement, it resends the segments. Figure 1.3 displays the Transport layer of the OSI reference model.

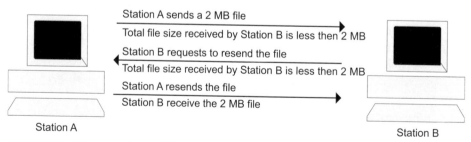

FIGURE 1.3   Transport layer of the OSI reference model.

The functions of the Transport layer are:

- Transfers data between hosts
- Performs error checking and recovery in case of loss of transmission
- Ensures data integrity by maintaining flow control
- Prevents buffer overflows for avoiding data loss
- Provides acknowledgement for successful transmissions
- Ensures complete data transfers using techniques such as CRC, windowing, and acknowledgements

Session Layer

This layer helps in establishing and terminating a connection. This helps certain applications to maintain session-based connections between client and servers. A *session* is a communication connection between the applications of two communicating hosts.

The functions of the Session layer are:

- Coordinates the interaction between two applications
- Establishes, maintains, and terminates the connection between two applications
- Manages more than one connection for one application
- Reconnects two hosts in case of an error
- Coordinates dialogs and data exchange
- Provides security to the system
- Determines the duration of a session
- Authenticates and identifies hosts
- Incorporates checkpoints in connection-oriented sessions
- Ensures methods to rectify errors occurring during data transfer
- Performs logon authentication
- Serves as an administrator for a session

Presentation Layer

This layer performs the coding and decoding function by converting the user-identifiable language into the machine language understood by the machines and devices. The user identifiable language is delivered to the Presentation layer from the Application layer. The Presentation layer also helps in the compression of data for quicker delivery, thus consuming less network bandwidth. This layer also provides security from the attackers by encrypting the data.

The Presentation layer performs these functions:

- Translates different application formats into standard format, and vice versa
- Transforms data with different structures into a standard format
- Formats data transfer syntax

- Performs encryption and decryption of data before transmitting across the network
- Performs data compression and decompression while transmitting data
- Performs conversion between graphic standards
- Uses Abstract Syntax Notation 1 (ASN.1), which is an international standard for representing data types and structures, as a standard data syntax

Application Layer

The Application layer is the upper layer of the OSI reference model. This layer handles high-level protocols, such as HTTP, FTP, and SMTP. These protocols are used by the Application layer to communicate with the Application layer on the destination system. In addition, this layer handles flow control and error recovery.

The functions of the Application layer are:

- Determines the source and destination of the intended communication
- Determines resources required for communication
- Synchronizes the communicating applications
- Determines authorization and authentication of communicating end users
- Supports different library functions
- Provides services for uniting the components of network applications, such as remote access, directory access, e-mail, and file transfer
- Provides measures required for detection of any error during the communication process
- Determines error recovery procedures
- Handles flow control during communication

Encapsulation of Data

According to Cisco, the process of placing data behind headers (and before trailers) of the data packet is called *encapsulation*. When a data packet travels from one layer to another, each layer creates a unique header and places the data received by the upper layer behind it. Table 1.2 shows the steps of data encapsulation within the OSI layers.

TABLE 1.2   Data Encapsulation

Layer	Functions
Application layer	Creates the application header and places the data created by the application after the header
Presentation layer	Creates the presentation header and places the data received from the Application layer after the header

TABLE 1.2 *(continued)*

Layer	Functions
Session layer	Creates the session header and places the data received from the Presentation layer after the header
Transport layer	Creates the transport header, places the data after the header, and passes this information to the layer after it
Network layer	Creates a header, places the data after the header, and passes it to Layer 2
Data-link layer	Creates a header, encapsulates the data after it, and also adds a trailer to end of the data and passes it to the first layer
Physical layer	Encodes a signal to transmit the data frame

Data encapsulation involves five conversion steps. These steps of data encapsulation in the TCP/IP Network model are:

1. Data is created. The user application is ready to send the data.
2. The transport header is created and data is placed behind the header.
3. The destination Network layer address is added to the data.
4. The Data-link address of the destination is added.
5. Transmission of the data bits occurs (the signal is encoding and sent).

Table 1.3 shows protocols, standards, and applications associated with each layer of the OSI model.

TABLE 1.3 OSI Reference Model and Associated Protocols, Standards, and Applications

Layer	Standards/Protocols/Applications
Application	• DNS • FTP and TFTP • BOOTP • SNMP and SMTP • MIME • NFS

(continued)

TABLE 1.3   *(continued)*

Layer	Standards/Protocols/Applications
	• FINGER
	• TELNET
	• NCP
	• APPC
	• AFP
	• SMB
Presentation	• PICT
	• TIFF
	• MIDI
	• MPEG
Session	• NetBIOS
	• NFS
	• RPC
	• Mail Slots
	• DNA SCP
	• Names Pipes
Transport	• TCP
	• SPX
	• NetBIOS/NetBEUI
	• ATP
	• ARP, RARP
	• NWLink
Network layer	• IGMP
	• IPX
	• NetBEUI
	• OSI
	• DDP
	• IP
	• ARP
	• RARP
	• ICMP
	• RIP
	• OSPF
	• IGMP
	• DECnet
	• X.25

TABLE 1.3   *(continued)*

Layer	Standards/Protocols/Applications
Data-link	• HDLC • SDLC • LAPB • PPP • ISDN • SLIP
Physical	• IEEE 802 • IEEE 802.2 • EIA/TIA-232 • EIA-530 • ISDN • RS232 • ATM

EVOLUTION OF TECHNOLOGY

Users today enjoy a free hand in bandwidth-intensive applications and knowledge sharing through the Internet, thanks to the advancement in technology over the years. A person working on a network in the 1970s and 80s did not have this freedom of bandwidth.

Things have changed manifold, and technology has grown from the 1-Mbps StarLAN to the 1000-Mbps Gigabit Ethernet. Local Area Network (LAN) technologies, such as Ethernet, Token Ring, and Fiber Distributed Data Interface (FDDI), have survived the research.

LAN ENVIRONMENTS

Earlier, most IT-enabled companies were having their own norms and rules for implementation of proprietary technology. After a decade, they agreed on globally accepted common standards for the network environment.

LAN Technologies

The term *topology* literally means "structure." Topology can be categorized in two parts: *physical topology* and *logical topology*. Physical topology is the physical structure of the network, while logical topology is how the terminal accesses the media. There are four basic types of topologies:

- Bus
- Star
- Ring
- Mesh

Bus

Bus topology involves the use of a single coaxial cable called a *trunk* or *backbone*. All the hosts of the network examine the data being transmitted and accept it only if the packet is addressed to them. Bus topology networks require that a terminator be installed at each endpoint of the bus to absorb the signal to avoid disturbance. The disadvantage of the bus topology is that the performance declines when a new workstation is added to the topology. Bus topology is not recommended for a large network environment because when the trunk cable goes down, the entire network goes down. Figure 1.4 displays a bus topology.

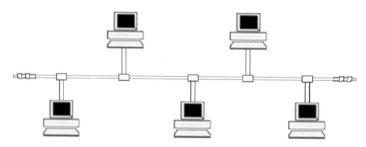

FIGURE 1.4   Bus topology.

Star

This topology is one of the most widely accepted topologies. It is easy to implement and troubleshoot the star topology. A central device, usually a hub or switch, is used as a central point of connectivity. An unshielded twisted or shielded twisted pair cable is used as a medium of connectivity. For connections to a hub, use an RJ45 connector. Figure 1.5 shows a Star topology.

Take an example of a company, ABCD Ltd., an organization that offers simple database management and document services. Most large organizations have functions that are complex, and therefore, such organizations prefer to outsource their database management services to ABCD Ltd., where there is a demand to

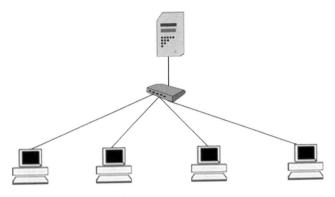

FIGURE 1.5 Star topology.

communicate with only a few remote terminals. In this case, a star network topology would be convenient since this network design allows easy troubleshooting features.

Ring

This topology involves connecting all computers in a network using a single cable representing a logical circle or ring. A central hub controls the connected computers. The data packets pass around the ring until they reach the destination computer. A break in the connecting cable can decrease the data transfer rate. There is no need to terminate the signal. A *token ring* network is based on ring topology. Figure 1.6 shows a ring topology.

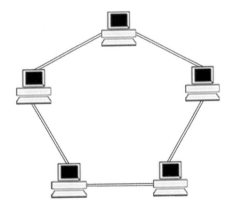

FIGURE 1.6 Ring topology.

Mesh

The mesh topology involves every computer having multiple possible paths with other computers on a network. This topology is fault tolerant while connecting one computer to another. It is tough to implement a true mesh topology. Figure 1.7 shows a Mesh topology.

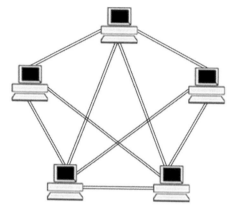

FIGURE 1.7   Mesh topology.

For example, to connect 100 computers in a network, 99 cables would need to be connected to *each* workstation. This makes mesh topology very expensive as well as difficult to implement. A true mesh requires at least three computers.

The number of network cables needed in a full meshed network can be calculated by using this simple formula:

$$N-1$$

where N is the number of nodes (computers or workstations) connected to the network.

LAN Protocols

Multiple network devices cannot communicate simultaneously, just as two people cannot talk effectively if they're talking simultaneously. Certain predefined rules are required for effective and smooth communication. A *protocol* is a set of rules that must be adhered to for effective communication between two entities. LAN entities use some protocols for communication. The major LAN technologies that a networking professional comes across on a daily basis are Ethernet, Token Ring, and FDDI.

Ethernet

Ethernet and Token Ring deploy access control methods when users transmit the data from the source to the destination. Access control methods set the rules that are followed to take control of the medium and to transmit data. The access method used by Ethernet is called *Carrier Sense Multiple Access with Collision Detection* (CSMA/CD).

CSMA/CD means that every device on a network segment must listen to the network before transmitting the data. The device checks for the presence of an electrical signal on the media, which indicates that somebody is already exchanging data with another device. If the device can detect an already present electrical signal on the cable, it must wait for a specific time and retry after that. If the number of retries exceeds the maximum limit, then the device can actually dispose of the frame. This can happen in situations when there are too many communicating devices on a network, which is a design issue in itself.

In companies that need a fast, inexpensive network that is also easy to administer, Ethernet would be the first choice. All hosts receive any signal broadcast over an Ethernet, and the concept of switching enables individual hosts to communicate directly.

Token Ring

The devices configured under the Token Ring network have more access controls on data transmission as compared to Ethernet. Token Ring uses a token (frame) to be shared among all the workstations. Workstations having the token have the right to transmit data. If any workstation does not have a token to send, it seizes the token, changes one bit of the token and appends the information, and sends it to the next workstation. This means that no other workstation can transmit until the information reaches the destination. Hence, there cannot be collisions on a Token Ring network. When the transmission is complete another workstation can start its transmission.

However, this does not mean that a device can keep the token for any amount of time. A token is passed among the devices, and has to spend an equal amount of time at each. The network device that connects the desktop in a Token Ring network is a *Multi-station Access Unit* (MAU). This unit also acts as a central hub in the network. The MAU takes care of the problem of loss of connectivity due to a break in the ring. It has internal circuitry to bypass the faulty device that is causing the break in the ring. A token ring can work at speeds of 4 Mbps and 16 Mbps.

Token Ring has a robust system of detecting the problem area and compensating for the network faults. To detect and correct network faults, Token Ring networks dedicate a station (active monitor) for monitoring frames, which are circling without being dealt with. When a sending device fails, its frame may continue to circle the ring thus preventing other stations from transmitting their own frames and consequently locking up the network. This monitor detects such continuously

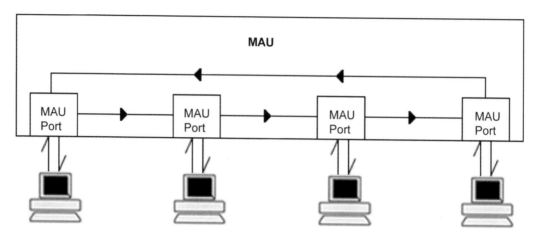

FIGURE 1.8   Token Ring network.

circulating frames from the ring, removes them, and generates a new token. Figure 1.8 shows a Token Ring network.

The Token Ring topology is preferred in the case of factory automation environments. In such a network scenario, applications must have predictable latency periods and robust network operation.

FDDI

Fiber Distributed Data Interface (FDDI) is a faster method of transmission than Ethernet cabling. FDDI was designed and developed by ANSI (American National Standards Institute) in 1980, which was later referred to ISO for further development and application in internetworks. FDDI has these requirements:

- A specified speed starting from 100 Mbps, which could scale up to gigabits
- Token-passing mechanism
- Failover ring, which gives reliability to the FDDI model
- Fiber-optic transmission media

FDDI employs two types of fiber cables: single mode and multimode. The difference between the two cables is the type of light used in them. While single mode uses a single laser light, the multimode fiber uses LEDs, which are introduced into the wire at multiple angles and thus can carry more traffic.

In case of organizations that need to transfer mission-critical applications from large computers to networks, FDDI would be preferred because this network topology offers benefits such as high speed and reliability.

LAN Devices

Devices commonly used in LANs include repeaters, hubs, LAN extenders, bridges, LAN switches, and routers. LAN devices, which are considered from the designing point of view, include repeaters, bridges, and switches.

Repeaters

A *repeater* is a LAN device that is used in networks connected with Thinnet and Thicknet. *Thicknet* and *Thinnet* are two common types of coaxial cables used in Ethernet LANs. Thicknet is 0.4 inches in diameter and has an impedance of 50 ohms. Thinnet is 0.2 inches in diameter with the same impedance as Thicknet. Although Thicknet was the original Ethernet wiring standard, Thinnet, which is cheaper, is more commonly used. The function of a repeater is to connect the wire segments so that their limitations due to distance do not hinder the process. It is an unintelligent device, which only amplifies the signal without understanding the source or destination of the data. It cannot segregate the collision or broadcast domains (which we will study in later sections). Repeaters exist on the Physical layer of the OSI reference model. In a single network, one can use a maximum of five repeaters in a series to extend the cable length. Figure 1.9 shows a repeater hub.

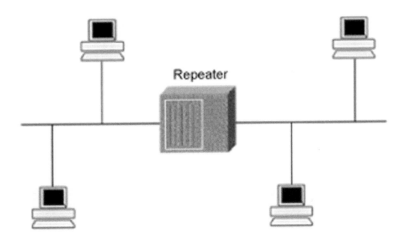

FIGURE 1.9 A repeater hub.

Bridges

The functions of repeaters are very limited and are not very useful in networks because they reside on the Physical layer. With the growth in the size of the network, it becomes necessary to partition the network into smaller groups of nodes to help isolate traffic and improve overall performance. One way to do this is to use a *bridge*.

Bridges exist on the Data-link layer of the OSI, which makes it possible to understand and work with MAC addresses. Bridges can watch the source and destination addresses and filter frames accordingly. This prevents the traffic from flooding segments, something that repeaters cannot do. Spanning Tree Protocol (STP) is one of the features of a bridge. In a pure transparent-bridging scenario, looping may occur if multiple active paths occur between two segments. STP is a protocol that modifies these multiple active paths into a single active path, putting the rest into blocking. The bridge intelligently sees that in case of link failures, the ST algorithm is re-run and an alternate path is made active. This feature is transparent to the user. Figure 1.10 shows how a bridge functions.

Organizations that have different types of cables in different workgroups use bridges. Segregating the workgroups using bridges would help in managing network traffic since inter-workgroup communication can occur only through bridges.

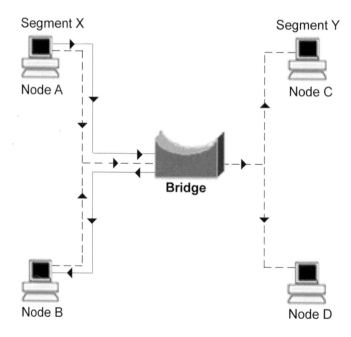

Node A

Node B

Node C

Node D

Segment X

Segment Y

Bridge

---- Messages from Segment X to Segment Y
——— Messages from Node A to Node B within the same segment.

FIGURE 1.10 Functioning of a bridge.

Switches

A *switch* is similar to a bridge but with some significant enhancements. First, a switch may have multiple ports, directing packets to several different segments and further partitioning and isolating network traffic similarly to a router.

Switches give desktop connectivity to the rest of the network and they form a vital part of 100BaseT and 1000BaseT networks. *100BaseT* and *1000BaseT* are the Ethernet standards that are popularly known as *Fast Ethernet* and *Gigabit Ethernet* respectively. They can work on Data-link as well as other upper layers, too. This is the reason they are more intelligent and understand MAC, IP, or IPX/SPX addresses (these address will be discussed later in the "LAN Addressing" section of this chapter). Apart from separating Collision Domains (discussed later in this chapter), a Cisco switch can also separate broadcast domains using Virtual Local Area Networks (VLANs). An enhanced feature of STP is available in the switch. This means that a switch can run STP on a per-VLAN basis. Figure 1.11 shows how a switch functions.

Switches segregate network traffic. This means that heavy-duty users such as programmers, CAD designers, and multimedia developers would be isolated from regular users. The light traffic producers would not be slowed down by the heavy user traffic. This separation would ensure that heavy users can be provided with high and dedicated throughput if needed.

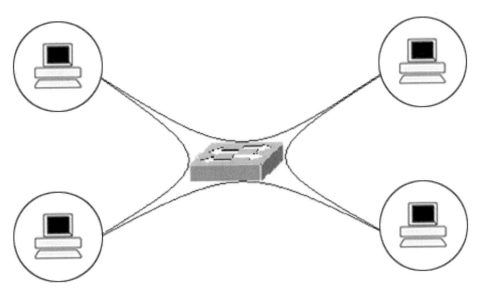

FIGURE 1.11   Functioning of a switch.

Look at Table 1.4 to understand how different network devices are placed on some of the layers of the OSI model. You will learn in detail about these devices in the further sections.

TABLE 1.4 Devices at OSI Layers

Layer	Devices
Network	• Routers • Layer 3 Switches • Bridge
Data-link	• Bridge • Layer 2 Switches • Network Interface Card
Physical	• Repeaters • Multiplexers • Hub • Transmission Cables

Traffic that Traverses LANs

There are three types of traffic that traverse LANs: *unicast*, *broadcast*, and *multicast* traffic. We will discuss each of them in detail.

Unicast Traffic

The network traffic consisting of packets traveling from a source to a destination is called unicast traffic. Each network device should have a unique MAC address to communicate on a LAN. If the source has specified a destination MAC address in the frame it sent, then the data packet with the MAC address is classified as unicast data. We see this in the example shown in Figure 1.12.

In Figure 1.12, when John with a NIC with the MAC address 00-0a-ba-fb-dy-8c, sends the data on the Ethernet LAN to the destination MAC address 00-0a-ba-9d-la-96, it will be received and processed only by Mike. Alen and Andy will simply ignore the frame and will not process it if they find that the destination MAC address does not match their own MAC addresses.

Broadcast Traffic

Unlike unicast traffic, *broadcast traffic* is peculiar in nature and should be rated carefully while designing networks. Broadcast traffic is a "free-for-all," which means all the devices on the Ethernet segment can process the frame and read the contents. This kind of traffic will not be ignored by the devices and will consume the CPU utilization of all the network devices. A broadcast frame sent by a network device

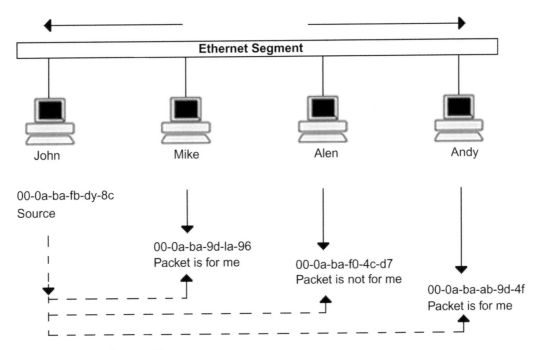

FIGURE 1.12   Unicast traffic.

has the destination address of FF-FF-FF-FF-FF-FF, which is reserved and is not allocated to any device in particular. Thus, all the devices process a broadcast frame.

This nature of broadcast traffic can sometimes bring a network to a halt if it's not controlled in a proper manner. In the next few chapters, we will learn how to control broadcast traffic. Consider the example of broadcast traffic illustrated in Figure 1.13. In Figure 1.13, each member of the network processed the broadcast frame.

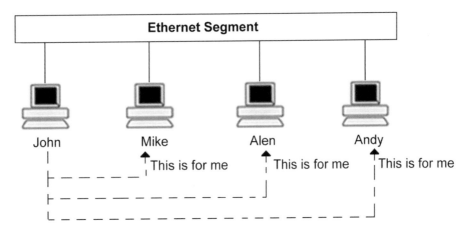

FIGURE 1.13   Broadcast traffic.

Directed Broadcast Traffic

A router uses this type of broadcast. A *directed broadcast* is an intentional broadcast created by the router for a particular host because routers do not forward broadcasts by default. An IP helper address is the command used to configure directed broadcasts. The *IP helper address* command is configured on the router's interface that receives the broadcasts. When this router receives broadcasts, it changes the destination broadcast address to a unicast address or a directed broadcast and forwards the packet.

Multicast Traffic

Multicast traffic is a refined form of broadcast traffic. All the multicast traffic is group specific. Only network devices that are part of some group and are configured for multicast traffic will process a multicast frame. This is used mainly in multimedia streaming for audio/video on public networks. For example, you visit an online music Web site that charges to let you listen to the online music live shows on the Internet. Those who pay the amount have to install the software, which will configure their Integrated Circuit for multicast traffic. Then they can log onto the Web site and enjoy the music show that will be delivered only to a registered user and not to everybody.

One more example could be the routing table updates sent by the router to update other routers. If a router broadcasts the updates, the traffic will be processed by all the devices on the LAN other than the routers. If the same router is multicast-compatible, it can send the traffic specifically to other routers only, without affecting other devices. This cuts down on the waste of bandwidth. Cisco uses 01-00-0c-cc-cc-cc as their multicast address to talk to other Cisco devices. Figure 1.14 shows an example in which the users are accessing a multimedia server.

Broadcast Domain

A broadcast domain is a network segment within which all the nodes are able to view and process the broadcast packets sent by any of them. If there were 50 workstations in a single broadcast domain and any of them sends a broadcast packet, the other 49 workstations would be able to process it. Figure 1.15 shows a broadcast domain.

In Figure 1.15, two LAN switches are connected to each other, supporting multiple workstations. These workstations are on an Ethernet network and use TCP/IP for communication. TCP/IP and other protocols are introduced later in this chapter and discussed throughout the book. Consider A as the source, which needs to transfer a 5-MB file to B, but the location of B is not known to A. To find out the address for B, A will send a broadcast packet on the segment, which will be processed by all the PCs including C and D. This packet will ask each PC to find

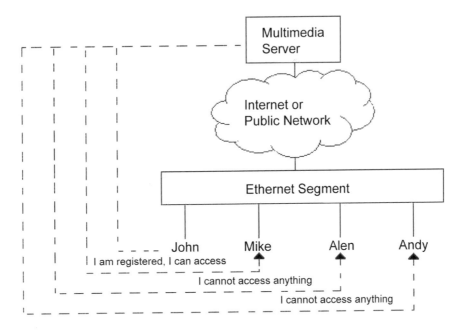

FIGURE 1.14   Multicast traffic.

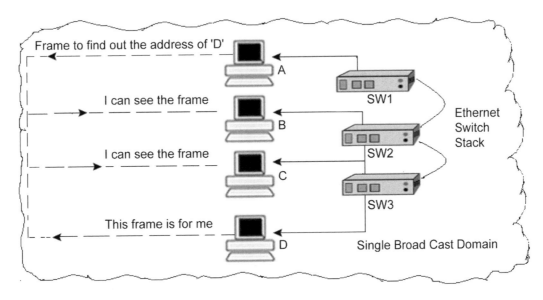

FIGURE 1.15   Broadcast domain.

out the address of B. To the query generated by A, only B will respond with its MAC address, and the actual data transfer can now take place. All the PCs that can process the broadcast packets generated by A are said to be in a single broadcast domain.

Collision Domain

A *collision domain* is a single area where two or more workstations are contending for the media, and collision might happen when they talk on the network at the same time. This collision occurs largely due to too many workstations connected in a flat architecture in which collision domains have not been set up. When there are too many workstations, they might not be able to sense every transmission on the entire network, and two or more workstations might therefore start to transfer data at the same time, potentially causing a collision to occur. You can look at the collisions on a switch by using the show interface fast ethernet command. Listing 1.1 shows the collisions after executing the show interface fast ethernet command.

LISTING 1.1   The show interface fastethernet command.

```
Switch# show int fastethernet 0/30
FastEthernet0/30 is up, line protocol is up
Hardware is Fast Ethernet, address is 0009.43b2.aa9e (bia
0009.43b2.aa9e)
MTU 1500 bytes, BW 100000 Kbit, DLY 100 usec,
reliability 255/255, txload 1/255, rxload 1/255
Encapsulation ARPA, loopback not set
Keepalive not set
Auto-duplex (Half), Auto Speed (100), 100BaseTX/FX
ARP type: ARPA, ARP Timeout 04:00:00
Last input never, output 00:00:00, output hang never
Last clearing of "show interface" counters never
Queueing strategy: fifo
Output queue 0/40, 0 drops; input queue 0/75, 0 drops
5 minute input rate 0 bits/sec, 0 packets/sec
5 minute output rate 5000 bits/sec, 8 packets/sec
140732337 packets input, 205260024 bytes
Received 583551 broadcasts, 0 runts, 0 giants, 0 throttles
791 input errors, 791 CRC, 0 frame, 181 overrun, 181 ignored
0 watchdog, 75460 multicast
0 input packets with dribble condition detected
1016424633 packets output, 292878449 bytes, 5973 underruns
1 output errors, 548651 collisions, 1 interface resets
```

```
0 babbles, 1 late collision, 315476 deferred
0 lost carrier, 0 no carrier
5973 output buffer failures, 0 output buffers swapped out
```

Table 1.5 shows the collisions on a LAN device.

TABLE 1.5 LAN Device Collisions

Device	Separate Collision Domain	Separate Broadcast Domain
Repeater	No	No
Bridge	Yes	No
Switch	Yes	Yes (Through VLANs)

LAN Addressing

There are two types of LAN addresses from the OSI model. These reside on Layers 2 and 3 of the model. The following discussion will help you to differentiate between Layer 2 addresses and Layer 3 addresses.

MAC Address

The Layer 2 or Data-link addresses have an important role in LAN addressing. The Data-link layer has a sublayer known as the Media Access Control (MAC), which is responsible for Layer 2 addressing. All the network devices have network interface cards (NICs) to communicate on the LAN. The NICs have a burned-in address, which means that it is hard coded to the NIC. This hard-coded address is a 48-bit-MAC address, which is unique for all NICs in the world.

This 48-bit-MAC address is expressed in hexadecimal form. Examples of MAC addresses are 00-06-5B-7E-43-52 and 000a.f306.0880 (for a router's Ethernet interface).

From the six octets in a MAC address, the first three octets are used by the NIC card manufacturers and are assigned to each of them by a standards body. This identification is also called the *Organizational Unique Identifier* (OUI). The remaining three octets uniquely identify the host. For example, in the MAC address 00-06-5B-7E-43-52, 00-06-5B is the OUI and 7E-43-52 is the host identifier for the manufacturer. Any two MAC addresses from the same manufacturer would never have the same last three octets.

Network Layer Address

Let us look at the Network layer address, which is used to build and address devices on LANs. Different protocols that exist on the Network layer, such as Internet Protocol (IP) and Internetwork Packet Exchange (IPX), use their own addressing methods to identify the host on the network. The TCP/IP Network layer address is known as an IP address, which is a 32-bit address divided into four octets. These four octets tell us two types of information about a node: the network ID and the host ID. Table 1.6 shows the various classes of IP addresses.

TABLE 1.6   Classification of IP Addresses

Class	IP Address Allocation
A	• The first octet is the network ID. • The remaining three octets comprise the host ID. • The address range is 1.0.0.0–126.0.0.0.
B	• The first two octets comprise the network ID. • The last two octets comprise the host ID. • The address range is 128.0.0.0–191.255.0.0.
C	• The first three octets comprise the network ID. • The fourth octet is the host ID. • The address range is 192.0.0.0–223.255.255.0.
D	• The address range is 224.0.0.0–239.255.255.255. • The address is not divided into network ID and host ID. • Used for multicast purposes.
E	• The address range is 240.0.0.0–255.255.255.0. • Reserved for experimental purposes.

WAN ENVIRONMENTS

A *Wide Area Network* (WAN) is a data communications network that covers a relatively large geographical area. Typically, a WAN consists of two or more LANs. The computers connected to WANs are often connected through public networks, such as the telephone system via a telecommunication provider (for example, AT&T and MCI). They can also be connected through leased lines or satellites. WANs are designed to operate over large geographical distances, allowing access over slow serial interfaces operating at much slower speeds than LANs with either full- or part-time connectivity. The largest WAN in existence is the Internet.

We will organize various technologies and methodologies into categories, discussing their roles and weighing their merits and demerits in modern WANs. A general WAN environment employs data links such as *Integrated Services Digital Network* (ISDN), *Frame Relay*, or *leased line*, that are provided by carrier services to access bandwidth over wide-area geographies. Routing protocols such as RIP, IGRP, EIGRP, and OSPF analyze and intelligently forward the data to the destination network with the knowledge of learned routes and maintained awareness of the network topology.

The Data-link layer protocols determine how data is encapsulated for transmission to remote sites, and they provide mechanisms for transferring the resultant frames. WANs use numerous types of devices such as routers, WAN switches, modems, Channel Service Unit/Data Service Unit (CSU/DSU) communication servers, and multiplexers. CSU/DSU servers convert network data into digital bipolar signals before transmitting them on a synchronous line. Routers segregate the individual LANs into separate broadcast and collision domains, effectively reducing overhead broadcast traffic over the WAN, and creating virtual private networks (VPNs). VPNs are divided into three types: *intranet*, *extranet*, and *remote access*.

A typical WAN setup will connect an organization's central site, usually a data center of the company's network (say Los Angeles), to the remote sites at separate locations as in Figure 1.16. The resources that remote offices, mobile users, and traveling individuals need to access are normally located here. This is the distribution site for all corporate network services. The remote site gains access to the corporate resources via WAN connections with the central site and, in some cases, may be a switching or gateway site for telecommuters or mobile users. An ISDN connection is favored for remote locations using external services, such as mail,

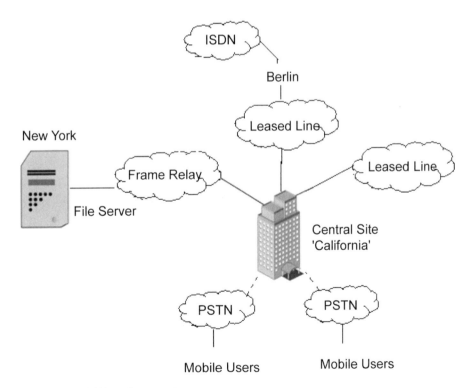

FIGURE 1.16 A WAN environment.

terminal emulation, and Frame Relay for accessing mail, databases, client/server applications, and video conferencing.

WANs usually carry different traffic types, such as voice, data, and video. The design selected must provide adequate capacity and transit times to meet the requirements of the organization. Among other considerations, the design must specify the topology of the connections between the various locations, along with availability, bandwidth, cost, application traffic, quality of service, reliability access control, and ease of management.

Connecting Broadcast Domains

Routers segregate the individual LANs into separate broadcast and collision domains because the router does not forward Address Resolution Protocol (ARP) tables and other Layer 2 information. Layer 3 devices have the job of connecting different broadcast domains over WANs and VLANs. They may be routers, switches, or other Layer 3 devices. The data packet with the address field of Layer 3 devices is dropped at the output port of the router, and unicast or multicast traffic is allowed. This is done by reading the address field of the Network layer.

Routers examine the Network layer address within each frame sent to them to make their forwarding decisions. The router strips the source MAC address from the frame and replaces it with its own MAC address when sending the frame outside. However, they maintain the original source MAC in buffered memory. When the appropriate data information is returned, it replaces the original MAC and sends it back. Thus, the different broadcast domains are connected over the WAN. Figure 1.17 shows connectivity within a broadcast domain.

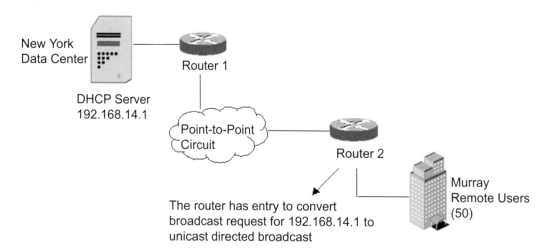

FIGURE 1.17   Connecting broadcast domains.

WAN Topologies

There are many options currently available for implementing WAN solutions; they differ in technology, speed, and costs. Knowledge of these technologies is an integral part of network design and evaluation. Let us have a look at the various WAN topologies.

Hub-spoke Topology

In a *hub-spoke* network topology, also known as a star topology, all the remote sites connect to the central node to form a network. This is a common design for packet-switched networks such as Frame Relay.

Figure 1.18 shows a Frame Relay network based on hub and spoke topology.

This topology is easy to manage and cost efficient because few *Permanent Virtual Circuits* (PVCs) connect the nodes to the central server. Bandwidth constraint is a major issue in this topology because all remote sites connect to each other through the central node or the core router, which has a limited bandwidth. In addition, there is only one possible path between each node and the central server. As a result, if network congestion occurs, the packet cannot be routed through an alternate path, and this results in a bottleneck since it is unable to scale to the increasing needs of the remote offices. If the central node fails, all the remote WAN sites connected to the central node go down.

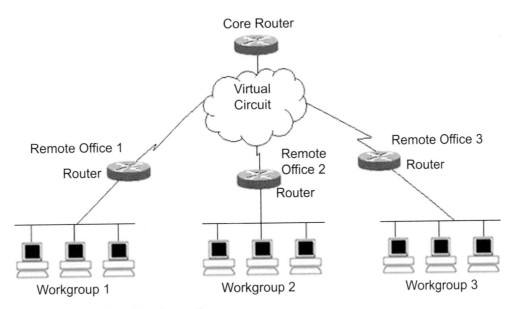

FIGURE 1.18   Hub and spoke topology.

Full Mesh Topology

In a *full mesh* network topology, each site is connected to all the other remote locations. Although a full mesh network allows all sites to communicate with each other, this topology requires a large number of virtual circuits for each router to connect to the switched network. Figure 1.19 displays a full mesh network.

This figure shows a full mesh Frame Relay network in which there is point-to-point connectivity among routers R1, R2, R3, and R4, with one PVC for each router. The advantage of a fully meshed network is the availability of a PVC for each network a router is connected to. This results in a high degree of network availability. The disadvantage of a fully meshed Frame Relay network is the high cost because a large number of PVCs connect routers to different networks.

A full mesh network is well suited for small Frame Relay networks. When the size of the network increases, management becomes a complex task. In addition, packets are replicated throughout the network because routers send broadcasts on each active logical connection. This results in additional bandwidth and CPU use by Data Terminal Equipment (DTE). Data Termination Equipment is a device that can transmit data in a digital format over a communications link.

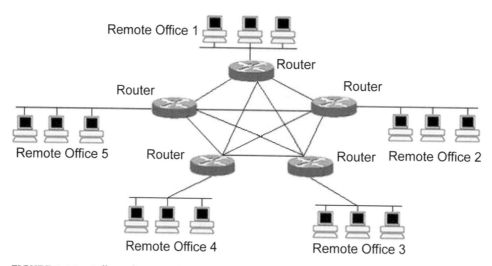

FIGURE 1.19 Full mesh network.

Partial Mesh Topology

In a *partial mesh* network topology environment, the number of virtual circuits is in between that in a full mesh and single point-to-point topology.

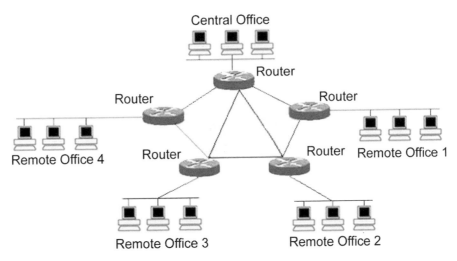

FIGURE 1.20 A partial mesh topology.

Figure 1.20 shows a partial mesh topology in which a full mesh topology and single point-to-point links coexist on the same network. Routers R3, R4, and R5 are part of the full mesh topology and are connected to each other. Routers R1 and R2 have only point-to-point connectivity. As a result, if R2 needs to connect to R3 or vice versa, both routers have to go through R2.

The advantage of this topology is that it allows the network to be divided into smaller subnets. Each subnet can be designed using a different topology, depending on the network traffic behavior and protocols.

WAN Technologies

If all data traffic occurs in an organization within a single location, a LAN setup fulfills the need of the organization. If data has to cross from building to building on a single site, the buildings can be interconnected with high-speed data links to form a campus LAN. However, if data has to be transferred between geographically dispersed locations, a WAN is needed to carry the data. The basic difference between WANs and LANs is that WANs require use of an external service offered by telecommunication companies such as AT&T, MCI, or others. We will discuss the various options and hardware available for these interconnections in later chapters.

From the technical design point of view, WANs can be divided into the following categories:

- Dedicated or leased lines
- Circuit switched

- Packet switched
- Cell switched

Dedicated or Leased Line

This service offers a point-to-point circuit between two offices. This circuit is set aside for the exclusive use of the leasing party, which pays a fee for the services. Examples of these types of services include, voice grade channels, digital services (for high-speed and low-error data transmissions), T-Carrier or E-carrier (for example, T1//E1, T3//E3, fractional T1/E1, DSL used to transport high-speed data and digitized voice), and SONET/SDH. They are best suited to long connect times and shorter distances. Leased lines give the highest bandwidth, maximum control, and maximum access.

Circuit Switched

When an organization establishes network connection points in various locations, it can use the public carrier's network to make temporary connections to each location, as needed, by using a switched circuit. They require a call setup and teardown; the initial signals are for negotiation and call setup and when the circuit has to release the connection, a teardown signal is sent signifying a call termination. This is similar to making a telephone call, picking up the handset from the cradle, and then dialing. As soon the call ends, the handset is replaced. The cost usually depends on the amount of time the network is used, or the volume of traffic passed through the network. Services include asynchronous Plain Old Telephone Service (POTS) ISDN (BR1/PRI), and Switched Multi-megabit Data Service (SMDS), essentially a switched form of SONET/SDH and T-Carriers or E- Carriers. ISDN is low-control shared bandwidth but delivers more bandwidth than asynchronous dialup, and is suitable for backup links and Dial-on Demand Routing (DDR). Analog dialups over the Public Switched Telephone Network (PSTN) are slower and less expensive than ISDN, and are best suited for short-term or on-demand access.

Packet-Switched Networks

Networks transport the data, forming virtual circuits. The subscriber has a local loop to the nearest location where the provider makes the service available. This is called the *Point-Of-Presence* (POP) of the service, which is essentially a dedicated circuit to the WAN provider's Central Office (CO). The service provider provides a permanent or switched virtual circuit to the destination. They are usually shared networks on the Telco (telephone company) cloud (the Telco cloud is the "cloud" representation of telephone companies on WAN diagrams). An internal addressing scheme is used in WANs over the Telco cloud, which is different than on LANs.

Packet switching is a data transmission technology that breaks information into smaller, more manageable groups called *packets*. As the information is being broken up into packets, overhead bits are added to the packet. Overhead bits include address bits and sequence bits. No actual transit route is specified in the information form to traverse from one point to the other. Before the first packet is sent through the network, it arrives at the switching equipment, which reads the address on the packet and determines the route that the packet will take. Using connection-oriented routing, the switching equipment creates a virtual circuit over which all other packets will travel. This circuit expires once the information has been sent. At the receiving end of the transmission, the sequence bits are read and the packets are reassembled into the proper sequence.

There are three stages to a virtual connection:

- Call setup
- Data transfer
- Call clearing

There are two types of virtual circuits: *Permanent Virtual Circuits* (PVC) and *Switched Virtual Circuits* (SVC). PVCs are end-to-end circuits, connections that are open for long periods of time and act in the same manner as private lines or leased lines. When a PVC is created, all subsequent calls do not require call setup. SVCs are end-to-end circuits that are set up at the initiation of a call and terminated when the call is ended. Users are charged based on the amount of usage of the network. Using the latest in network technology, services include Frame Relay, Asynchronous Transfer Mode (ATM), and Cellular Digital Packet Networks (CDPD-wireless data transmission), as well as older X.25. X.25 delivers low bandwidth, low control, and is invariably cost effective; its advantage is its high reliability. Frame Relay delivers medium control, shares bandwidth, and is a lower-cost alternative for backbones and branch sites. It is best suited for longer distances and connect times.

Cell-switched Networks

Cell-switched networks are similar to packet-switched networks, but they differ in that there are no variable length packets or frames in cell-switched networks, but fixed-length packet cells instead. ATM and SMDS networks are examples of cell-switched networks.

Most of the WAN technologies fit into one of the aforementioned categories. These subcategories can be collectively grouped as leased line networks. The different WAN link options are:

- Analog (Asynchronous) dialup
- Integrated Services Digital Network (ISDN)
- Leased line
- X.25
- Frame relay
- ATM

Analog (Asynchronous) Dialup

Analog dialup is a WAN option used for low-volume traffic, and as such is good for sporadic usage. This is because using an analog modem and voice telephone lines provides slow transmission, with speeds ranging from 33 to 56 Kbps. Dialups are in POTS, and a local loop is formed using copper cables from the customer's premises to a PSTN network. Modems and analog lines are simple, easily available, and incur low implementation costs. The demerits are delay (jitter), low data rate, and relatively long connection times. Some Small Office/Home Office (SOHOs) can survive on POTS. SOHO refers to the small or home office environments and the business culture that surrounds them. Figure 1.21 illustrates a simple modem-to-modem connection (an analog dialup) through a WAN.

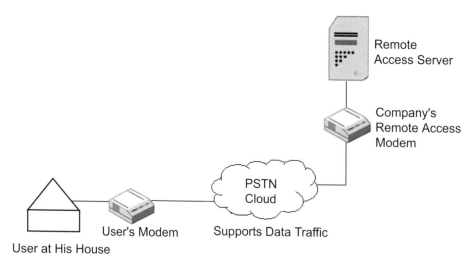

FIGURE 1.21 An analog dialup.

Integrated Services Digital Network (ISDN)

ISDN evolved from telephony to provide end-to-end digital connectivity to support a wide range of services. ISDN is an international communications standard for sending voice, video, and data over digital telephone lines or normal telephone wires. Users have access to ISDN by a limited set of standard multipurpose customer interfaces. ISDN uses *Time Division Multiplexed* (TDM) digital connection in a local loop. The connection uses *Bearer* channels (B) for carrying voice or data and a signaling, *Delta* channel (D) for call set-up and other purposes. ISDN is offered in two forms: *Primary Rate Interface* (PRI) and *Basic Rate Interface* (BRI). BRI offers lower data rates (up to 128 Kbps) and is suitable for residential and small business users. It has 2 B (64 Kbps each) and one D channel (16 Kbps). PRI is used by large businesses because it offers high data transfer rates (up to 2.048 Mbps). The method employed by ISDN to connect devices of Customer Premises Equipment (CPE) and Telco depends upon the reference point of the ISDN. We will discuss the reference point later in this section. Figure 1.22 shows an ISDN setup.

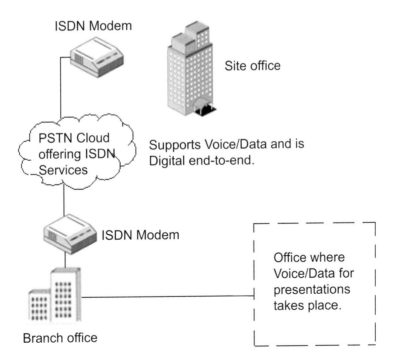

FIGURE 1.22 An ISDN setup.

In North America, the PRI is offered as 23 B channels and one D channel, while in Europe and Asia there are 30 B channels and one D channel. Multiple B channels can be connected between two end points. This connectivity allows for video conferencing and high bandwidth data connections with no latency or jitter. An ISDN *Terminal Adapter* (TA) is a device used to connect ISDN BRI connections to other interfaces. A TA is essentially an ISDN modem. The tariff is based on distance and capacity use of bearer channels. ISDN equipment is divided into functional groups and the ISDN reference points specify the communication protocols between these groups:

Terminal Equipment type 1 (TE1): Refers to specialized ISDN terminals, such as digital telephone and integrated voice/data terminals.

Terminal Equipment type 2 (TE2): Refers to non-ISDN terminals, such as terminals with RS-232 physical interfaces, host computers with an X.25 interface, and analog modems.

Terminal Adaptor (TA): Adapts TE2 devices and other equipment such as Ethernet interfaces to allow communication with ISDN networks.

Network Termination type 1 (NT1): Belongs to OSI Layer1, and is associated with physical and electrical termination of the ISDN on the user's premises.

Network Termination type 2 (NT2): Belongs to OSI Layers 2 and 3. These are intelligent devices functioning as switches or concentrators.

Table 1.7 lists the reference points in an ISDN.

TABLE 1.7 Reference Points in an ISDN

Reference Point	Details
R	Reference point between a non-ISDN compatible TE and a TA
S	Reference link between TE or TA and NT equipment
T	Reference point between user-switching equipment and NT1
U	Reference point between NT equipment and the Local Exchange
V	Interface point between the line termination and the exchange termination

Leased Line

Leased lines are used for building WANs and giving permanent dedicated capacity. These dedicated, pre-established point-to-point connections offer speed up to a few Gbps. If connecting multiple sites, these connections are costly as compared to the popular Frame Relay. The advantages of these connections are that they are permanent, dedicated, and high speed, and that they experience negligible delay or jitter. Figure 1.23 shows a leased line.

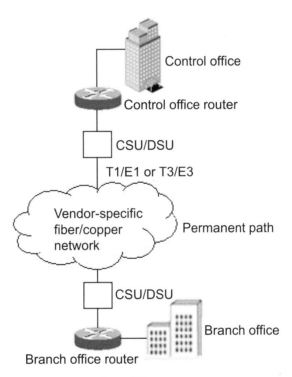

FIGURE 1.23   A leased line.

A DTE serial port is required on the router and Channel Service Unit/Data Service Unit (CSU/DSU) to complete the circuit at the customer premises. A CSU/DSU is a digital-interface device, or sometimes two separate digital devices, that adapt the physical interface on a DTE device (such as a terminal) to the interface of a DCE device (such as a switch) in a switched-carrier network. DTE (Data Terminal Equipment) is end devices and DCE (Data Communications Equipment) is intermediate devices that DTE end devices communicate through. Leased line tariffs depend upon distance and capacity. *V.35* cables are used for connectivity.

They are also called *DCE-DTE* cables or *Back-to-Back cables*. Table 1.8 shows the different link types, signal standards, and bandwidths.

TABLE 1.8   Line Types, Signal Standards, and Bandwidth

Link Type	Signal Standard	Bandwidth
56	DS0	56 Kbps
64	DS0	64 Kbps
T1	DS1	1.544 Mbps
E1	ZM	2.048 Mbps
T3	DS3	44.736 Mbps
E3	M3	34.064 Mbps
OC1	SONET	54.184 Mbps
OC3	SONET	155.54 Mbps
OC48	SONET	2.43 Gbps

X.25

X.25 is now an internationally accepted transmission protocol for connecting various types of terminals and computers to packet-switched data networks. Although it operates at much slower speeds than many of the others, ranging from 48 Kbps to 2 Mbps, it has excellent reliability built in. It employs the LAPB protocol for connecting the various terminals in the network. Most large organizations use X.25 to connect to remote servers to synchronize their billing database. Advantages of X.25 are:

■ X.25 Packet Switched networks allow remote devices to communicate with each other across high-speed digital links without the expense of individual leased lines. For example, you can have a host connected at 56 Kbps communicating with numerous remote sites connected with cheaper 19.2 Kbps lines.
■ X.25 provides high quality digital networking at low cost. This proves to be economical when multiple parties share the same infrastructure.
■ X.25 is very stable as there are practically no data errors on modern X.25 networks.

Frame Relay

Frame Relay (FR) is a packet-switched data transport technology that facilitates the transportation (relaying) of variable length units of data called *frames*. It was developed as an alternative means of data transport that could eventually replace the packet-switching standards, including X.25. It employs two circuit types: *Permanent Virtual Circuit* (PVC) or *Switched Virtual Circuit* (SVC). PVC is a permanent, predefined path in the Telco cloud, but SVCs are temporary paths in the Telco cloud. The data frame can take any of the paths to cross the SVC. FR can have either type or only one type as a switched path to the destination network.

The Frame Relay standard specifies only the first two layers of the OSI standard, Physical and Data-link. The tariff costs are based on the capacity of the link. The tariff charges vary rather interestingly as the subscriber can opt for a *Committed Information Rate* (CIR), a *minimum guaranteed bandwidth*, and a *Burst Rate* (BR) value, which is much higher than CIR. With a threshold value of CIR, subscribers can send data over the shared network at the speeds mentioned for CIR, but if the shared network has more available capacity, the traffic can burst up to speeds mentioned for the BR. The service provider usually charges for the CIR value. It gives permanent variable connections with bit rates exceeding 4 Gbps.

Asynchronous Transfer Mode (ATM)

ATM is based on a fixed-length packet called *cell technology*, rather than on frame-based architectures. A cell has a data payload of 48 bytes and a 5-byte header. These fixed-length cells are intolerant to delay or jitter so they are useful for video and voice transmission over data communication channels, as no waiting period is required for large data packets.

True integration of voice, data, and video in one technology has been possible with ATM. A very low latency and jitter rate at much higher bandwidths has made the *Quality of Service* (QoS) of ATMs attractive. ATM uses both types of virtual circuits, PVC and SVC. It has a variable capacity with speeds reaching above 156 Mbps.

New Trends in WAN Technologies

WAN technologies function at three layers of the OSI reference model, Physical, Data-link, and Network. Let us discuss various protocols and technologies used in Wide-Area Network (WAN) environments.

Cable Modems

Enhanced cable modems enable two-way, high-speed data transmission using the same coaxial lines that transmit cable television. Cable modems provide an always-on connection and a simple installation. This technology is usually meant for *Metropolitan Area Networks* (MANs), but offers potential for WANs by multiplexing signals over high capacity links at the service provider end. It offers speeds in multiples of T1/E1.

Digital Subscriber Line

Digital Subscriber Line (DSL) technology allows the Telco to provide high-speed, always-on network services to the subscribers, utilizing existing local loop copper lines. DSL service is considered broadband, as opposed to the baseband service for typical LANs. Different varieties of DSL provide different bandwidths with capabilities exceeding those of a T1 or E1 leased line. For an optimal connection, the local loop must be less than 3.5 kilometers (2.2 miles). DSL technology is available in many forms, including:

- Asymmetric DSL (ADSL)
- High Bit Rate DSL (HDSL)
- Symmetric DSL (SDSL)
- ISDN (like) DSL (IDSL)

Cable Modems and DSL technologies are considered appropriate for MANs, but Telcos can combine these with other WAN technologies. A primary example of the use of DSL is Internet connectivity. Table 1.9 compares the different WAN technologies.

TABLE 1.9 WAN Technologies Comparison

Category	Analog Dialup	ISDN	Leased Line	X.25	Frame Relay	ATM
Maximum Bit Rate	33 to 56 Kbps	64 Kbps to 2.048 Mbps	<2.5 Gbps	<48 Kbps	<4 Gbps	<156 Mbps
Charges based on	Time and distance	Distance and capacity	Distance and capacity	Volume	Capacity	Capacity
Connection Type	Dialup	Dialup	Dedicated	Switched, fixed	Virtual switched or Permanent	Cell switched, SVC or PVC
Speed	Slow	Moderately fast	Very fast	Moderately slow	Very fast	Fast
Bandwidth	Limited/ shared	Shared	Full	Shared	Shared	Shared
Control	Low	Low	High	Low	Medium	Medium
General Characteristics and Use	SOHO, backup	Higher bandwidth than dialup, DDR	High-speed. Always on. High-cost enterprise	High reliability	High speed. Bandwidth on demand. Point to Multipoint. Medium cost enterprise	Voice, video, and data integration

WAN Connectors

In this section we will discuss the different WAN connectors and the practical aspects of a WAN environment. Different WAN connectors are as follows:

EIA/TIA-232: Common Physical layer interface, developed by Electronics Industries Association (EIA) and Telecommunications Industries Association (TIA). This type of connector has speeds of up to 64 Kbps. It was formerly called "RS-232," and referred to as serial ports with DB9 connectors.

EIA/TIA-449: Popular Physical layer interface, developed by EIA and TIA. It is a faster version of EIA/TIA-232, with speeds of up to 2 Mbps. It supports longer cable lengths than EIA/TIA-232.

EIA/TIA-612/613: *High Speed Serial Interface* (HSSI), with speeds of up to 52 Mbps. It provides an interface to T3, E3, and STS-1.

V.35: ITU-T standard, Physical layer: This is a synchronous protocol between DTE and a packet network connected to DCE, supporting speeds of up to 48 Kbps.

X.21: ITU-T standard for serial communication over synchronous digital lines.

G.703: ITU-T electrical and mechanical specification for connection between Telco DCE devices and customer DTE devices, with speeds equaling E1 data rates.

WAN Protocols

Protocols are like languages, in which discipline and control are maintained for verbal exchanges between individuals. Similarly, WAN protocols used over WAN topologies lay down a blueprint for communications between two or more devices. It is important to have the knowledge of Cisco-supported WAN protocols for making better design options. The types of WAN options are:

- Synchronous Data Link Control (SDLC)
- High-level Data Link Control (HDLC)
- Point-to-Point Protocol (PPP)
- Integrated Services Digital Network (ISDN)
- Frame Relay
- Routed protocols
- Routing protocols
- Encapsulation protocols

Synchronous Data Link Control (SDLC)

The SDLC procedure transports SNA information across the network. SDLC is IBM's specific implementation of the ISO standard High-level Data Link Control (HDLC).

SDLC features include support for point-to-point and point-to-multipoint links as well as full-duplex media. SDLC is also considered a very reliable protocol. It uses the terminology of primary and secondary station network node types.

High-level Data Link Control (HDLC)

HDLC is the default Cisco Data-link layer protocol used for synchronous serial leased lines. For this protocol, one requires Cisco routers on both ends. HDLC is the ISO-standardized protocol derived from SDLC, but Cisco has modified it and made a proprietary protocol based on it, so different vendor platforms may not be compatible. HDLC supports both point-to-point and multipoint configurations.

Point-to-Point Protocol (PPP)

PPP is a Data-link layer protocol, used for both asynchronous and synchronous media connection. This protocol uses Network Control Protocol (NCP) for the establishment and configuration of different Network layer protocols (IP, IPX) and Link Control Protocol (LCP) for establishing, configuring, and testing the Data-link connection.

LCP is designed to agree upon the encapsulation format options, handle varying limits on sizes of packets, detect looped-back links and other common misconfiguration errors, and terminate the link. With Multilink PPP, using synchronous (ISDN) and PPP encapsulation, Cisco supports multiple connections over one physical circuit. There is support for authentication through Password Authentication Protocol (PAP) and Challenge Handshake Authentication Protocol (CHAP).

Integrated Services Digital Network (ISDN)

ISDN is a suite of communication protocols (ITU-T standards) framed by telephone companies that allow the companies to deliver data and voice. ISDN uses two types of signaling. For communication between the customer premises and the central office, ISDN uses *Digital Subscriber Signaling System 1* (DSS1). DSS1 sets the format of the data that goes in the D-channel and message formats used for connection establishment, maintenance, and termination. For signaling between network components, *Signaling System 7* (SS7) is used. SS7 defines communication protocols and formats similar to DSS1. The connectivity between these two signaling protocols needs a convergent function. This function maps the LAPD (an ISDN protocol) frame from the customer part to the SS7 signaling format of the network part. DDR is also a very good feature of ISDN; it is used for passing traffic by initiating a WAN link and for configuring WAN backups.

ISDN Protocols

ISDN protocols are broadly classified into four types:

Link Access Protocol on the D-channel (LAPD): LAPD is a bit-oriented protocol on the Data-link layer of the OSI reference model. Its prime function is to ensure the error-free transmission of bits on the Physical layer. LAPD works in the Asynchronous Balanced Mode (ABM). This mode is totally balanced, with no master-slave relationship. Each station may initialize, supervise, recover from errors, and send frames at any time. The protocol treats the DTE and DCE as equals.

E series protocols: Offers use of ISDN over the existing telephone network

I series protocols: Covers concepts, terminologies, and services

Q series protocols: Deals with switching and signaling

Frame Relay

Frame Relay is a standard for internetworking, which provides a fast and efficient method of transmitting information from a user device to bridges and routers. The Frame Relay standard uses a frame with a structure similar to that of LAPD. It also contains congestion and status signals, which the network sends to the user. The frame header is altered slightly to contain the Data Link Connection Identifier (DLCI) and congestion bits in place of the normal address and control fields. DLCI is a 2-byte user-specified Frame Relay header field containing the destination address of the frame.

Frame Relay specifies Data-link specifications (LAPF headers), PVC Management, Local Management Interface (LMI), SVC Signaling, and multiprotocol encapsulation.

Virtual Circuits

The Frame Relay frame is transmitted to its destination by way of *virtual circuits* (logical paths from an originating point in the network) to a destination point. Virtual circuits may be permanent (PVCs) or switched (SVCs). PVCs are set up for a dedicated point-to-point connection; SVCs are set up on a call-by-call basis. Most Frame Relay configurations use PVCs.

Frame Relay is used increasingly nowadays for data transmission instead of X.25 networks and dedicated lines due to its high reliability and operational speed. As virtual circuits consume bandwidth only when they transport data, many virtual circuits can exist simultaneously across a given transmission line. Besides, each device can use more of the bandwidth as necessary, and thus operate at higher speeds.

The superior reliability of communication lines and error-handling sophistication at end stations today allows the Frame Relay standard to discard erroneous frames and thus eliminate time-consuming error-handling processing.

Local Management Interface (LMI)

The LMI protocol was developed in 1990 by a consortium of technology companies led by Cisco Systems. LMI is a signaling protocol used between DTE (router) and DCE (FR switch) that ensure that the access link between the FR switch and the router over the leased line is working. It notifies every other (FR switch and router) of status and problems on the link.

Multiprotocol Encapsulation

Multiprotocol encapsulation provides a flexible method for carrying multiple protocols on a given Frame Relay connection. This is useful when there is a need to multiplex/demultiplex across a single Frame Relay connection.

In an encapsulation technique a packet of data called a *Protocol Data Unit* (PDU) is taken from the LAN and prefixed by a header that defines the encapsulated LAN protocol. This is then converted into one or more Frame Relay packets. Frame Relay transports X.25, SNA, and LAN traffic with equal ease. Some examples of encapsulation protocols are CISCO, ANSI, and q933a.

Link Access Procedure Frame (LAPF)

Link Access Procedure Frame Bearer Services creates the Frame Relay header and trailer. The header includes these bits:

Data Link Connection Identifier (DLCI): 10-bit field representing the address of the frame and corresponds to a VC.

Forward Explicit Congestion Notification (FECN): Header bit transmitted by the source terminal requesting that the destination terminal slow down its requests for data, as congestion is occurring in the same direction as the frame.

Backward Explicit Congestion Notification (BECN): Header bit transmitted by the destination terminal requesting that the source terminal send data more slowly, as congestion is occurring in the opposite direction as the frame.

Discard Eligibility (DE): Provides the network with a signal to determine which frames to discard. When there is congestion on the line, the network must decide which frames to discard in order to free the line. The network discards frames with a DE value of 1 before discarding other frames.

Routed Protocols

Routed protocols are the network protocols that define the logical addressing structure and are responsible for communication at Layer 3. End systems such as terminals and printers use routed protocols for communication among themselves. The examples of routed protocols include IP in a TCP/IP stack and Internetwork Packet Exchange (IPX).

Internet Protocol (IP)

The Defense Advance Research Projects Agency (DARPA) developed the TCP/IP protocol suite to interconnect various Defense Department computer networks. The IP is a Network layer (Layer 3) protocol of the TCP/IP suite that contains addressing information and some control information that enables packets to be routed. IP is documented in RFC 791 and is the primary Network layer protocol in the Internet protocol (TCP/IP) suite. Along with the Transmission Control Protocol (TCP), IP is the heart of the Internet protocols. IP primarily performs two functions:

- Provides connectionless, best-effort delivery of datagrams through an internetwork
- Provides fragmentation and reassembly of datagrams to support data links with different *Maximum-Transmission Unit* (MTU) sizes

The TCP/IP protocol stack maps to the OSI reference model. The upper three layers, namely Application, Presentation, and Session are combined into a single layer called Application. The Transport layer maps to the host-to-host layer in the TCP/IP Protocol, and the Network layer is called the Internet layer. The Datalink and Physical Layers are concatenated into one layer called Network Access. TCP is a connection-oriented reliable protocol, and IP is the logical addressing protocol on which this protocol stack is built. IP routed protocols include Internet Control Message Protocol (ICMP), Internet Protocol v4, TCP, UDP, and Telnet.

Internetwork Packet Exchange (IPX)

Novell's Internetwork Packet Exchange (IPX) is the communication protocol for Netware OS. Its implementation on a Cisco router can be done along with TCP/IP in IOS® with multiprotocol support.

Routing Protocols

Routing protocols direct routed protocols through a network. When intermediate systems communicate among themselves, routing protocols are necessary. Routing protocols are algorithms on a router that decide the best path to the destination and maintain awareness about the network topology. Using these algorithms, the router directs the data information for the best match in the routing table entry and forwards it out of the output to the next hop (router).

Some examples of routing protocols used in Cisco environments are: Routing Information Protocol (RIP), Interior Gateway Routing Protocol (IGRP), Enhanced Interior Gateway Routing Protocol (EIGRP), Open Source Path First (OSPF),

Border Gateway Protocol (BGP), and Intermediate System-to-Intermediate System (IS-IS). Table 1.10 lists the differences between routed and routing protocols.

TABLE1.10 Routed and Routing Protocols

Routed Protocols	Routing Protocols
Define addressing and Layer 3 header in a packet that is forwarded by a router	Define the process of exchanging topology data such that the routers know how to forward the data
IP IPX, OSI, DECNET, AppleTalk, VINES	RIP, IGRP, EIGRP OSPF, NLSP, RTMP VTP, IS-IS

Encapsulation Protocols

A WAN connection uses a Data-link layer protocol to encapsulate traffic when data is traversing over a WAN link. The Data-link layer attaches a header on the frame from the Network layer data for necessary checks and controls. The choice of encapsulation protocols depends on the WAN technology and the equipment. Most of the framing is based on the HDLC standard, but it has been modified by different vendors for delivering proprietary encapsulation protocols. Table 1.11 shows the encapsulation protocols for different WAN technologies.

TABLE 1.11 Encapsulation Protocols

Encapsulation Protocol	WAN Technology
High-level Data Link Control (HDLC)	Synchronous leased lines
Point-to-Point Protocol (PPP)	Dialup connections
Link Access Protocol Balanced (LAPB)	X.25
Link Access Control Frame (LAPF)	Frame relay (CISCO and IETF)
Link Access Protocol—Channel D (LAPD)	ISDN D channel

WAN Devices

Devices commonly used in WANs include routers, public switches, X.25 switches, ATM switches, modems, servers, and multiplexers. Some of the WAN devices are routers, public switches, and Layer 3 devices.

Routers

Routers are relay devices that forward data packets to logical destinations. A router operates at the Network layer of the OSI model and performs routing and switching functions. The routing function is a process of learning and maintaining awareness of the network topology. A switching function is the actual movement of data traffic from an input to the output. Routers operate on the concept of logical addressing (such as IP addressing) with the capability of making intelligent decisions (with the help of routing protocols) concerning the flow of information.

Routers examine frames addressed to them by scanning the network address within each frame to make their forwarding decisions. A router then examines its internal routing table (or database) to determine where to send the packet. The router then sends the packet out on the interface that has been specified for the next hop. Cisco has many routers based on its proprietary Internetwork Operating System (IOS) with advanced built-in electronics. Some of the examples are shown in Table 1.12.

TABLE 1.12 Different Routers and their Functions

Router Series	Functions
7200 series	Top of the line, supporting an enterprise-class range of interfaces. This series is used by Telcos and ISPs.
4000 series	Access routers are similar to the 7200 but are better for large regional office environments. They also support PRI and T1/E1.
3600 series	Used for corporate central sites. The number of network module slots varies from 2 to 6, according to the model. It is a RISC-based router and is capable of handling enterprise-level routing traffic.
2600 series	These come in several fixed configurations that can include one network module and up to two WAN modules.
2500 series	Offers BRI, serial, and other WAN interfaces and is also fixed configuration.
1700 series	A physically small device that's ideal for a small office, the 1700 has an Ethernet port and two modular slots for WAN connections. Select models come with a voice WAN slot.
1600 & 800	Routers for SOHO and telecommuter use series.

Public Switches

Public switches are used in WANs for transferring data over public circuits. Public switches include X.25, ATM, and Frame Relay switches.

X.25 and frame relay switches: Using packet switching technologies, these WAN switches transfer data over public (shared) data circuits leased via the Telco.

ATM switches: These WAN switches facilitate high-speed cell-switching between WAN and LANs. Table 1.13 shows the equipment range available from Cisco.

TABLE 1.13 Cisco ATM Switches

Series	Explanation
Cisco IGX series	ATM-based WAN switches connect to public services. There are a maximum of 32 slots, which can house ATM, Frame-relay, or Voice modules.
Cisco MGX series	These have edge concentrators. There is a cost-effective narrowband multiservice option, and the unit scales from DS0 to OC-48c/STM-16 speeds.
Cisco BPX series	Large-scale advanced ATM switch for service providers and large enterprise applications.
Cisco LS1010	Layer 3 enhanced ATM switch seamlessly integrating wire-Speed Layer 3 switching and ATM switching into a single chassis.

Layer 3 Devices

Layer 3 devices are used in a WAN environment for data transfer over a physical circuit. Different Layer 3 devices are:

Multiplexers (MUX): They combine (*multiplex*) digital signals over one channel to send data out on the physical circuit. Different signals are sent over one communication channel with use of this equipment.

Communication servers: These are remote access servers for dial in or dial out from diverse locations over the WAN or LAN. Generally, telecommuters or remote branches dial into it remotely, and access the remote services through it. Products for this category of device are the AS5300 series, AS5400 series, and AS 5800 series.

Cisco AS5300 series: Mixed-media environments with ISDN, synchronous serial, and asynchronous functionality, as well as multiple T1/E1 and PRI.

WAN Addressing

WAN addressing is based on a rather different sort of convention.

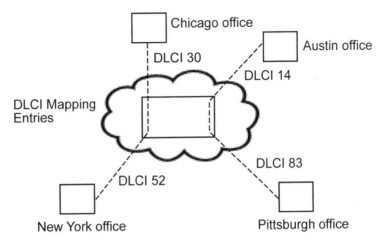

FIGURE 1.24   Frame relay addressing between four locations.

Frame Relay Addressing

Consider four locations: Chicago, Austin, New York, and Pittsburgh. As shown in Figure 1.24, there is a PVC between Chicago and New York and another between Austin and Pittsburgh.

Chicago uses DLCI 30 to refer to its PVC with New York, whereas New York refers to its PVC with Chicago as DLCI 52. Similarly, Austin uses DLCI 14 to refer to its PVC with Pittsburgh, whereas Pittsburgh refers to its PVC with Austin as DLCI 83. The DLCI can be a maximum of 10 bits; hence 1024 addresses can be present in the Frame Relay cloud, addressing devices. The available addresses from these depend upon the LMI type used. The addresses not available in the network are for vendors, and LMI messages and multicasting. With Frame Relay offering means for multiplexing many logical data conversations, the Telco switching equipment (FR Switch—DCE) first constructs a table mapping DLCI values to outbound ports. When a frame is received, the switching device finds the connection identifier and delivers the frame to the respective outbound port. The complete path to the destination is established before the first frame is sent.

SUMMARY

In this chapter, we learned about the building blocks of a network and its evolution. We also reminded you about the OSI reference model and LAN and WAN networks. In the next chapter, we move onto enterprise networks and their classifications.

POINTS TO REMEMBER

- The prime function of the Physical layer is to interact with the transmission media and to put the data on the media in the form of bits.
- Cables, and devices such as hubs and repeaters reside on the Physical layer.
- The Network Interface Card (NIC) is a part of Layer1, but functions at Layer 2.
- The Data-link layer has two sublayers, Media Access Control (MAC) and Logical Link Control (LLC).
- The MAC sublayer provides an address for each device on the network, while the LLC establishes and maintains links between the communicating devices.
- The Network layer defines a mechanism to deliver packets from one network to another, which forms an internetwork.
- The Transport layer divides the data into segments, which are easier to deliver and track.
- TCP is a connection-oriented feature of the Transport layer.
- UDP adds a connectionless feature to the Transport layer.
- The Session layer helps in establishing and terminating a connection.
- The Presentation layer performs the coding and decoding function by converting the user-identifiable language into machine language.
- The Application layer handles high-level protocols such as HTTP, FTP, and SMTP.
- The major LAN protocols are Ethernet, Token Ring, and FDDI.
- All network devices communicate with the NIC, which is uniquely identified by a 48-bit MAC address in hexadecimal form.
- The access method used by the Ethernet is called Carrier Sense Multiple Access with Collision Detection (CSMA/CD).
- In a Token Ring network, a token has to pass among the devices, spending an equal amount of time at each node in the network.
- The main network device that connects the computers in a Token Ring network is a Multistation Access Unit (MAU).
- Fiber Distributed Data Interface (FDDI) is a faster method of transmission than Ethernet cabling.
- The difference between single mode and multimode fiber optic cables is the type of light used in them.
- New additions to WAN technologies include cable modems and Digital Subscriber Line (DSL).
- A cable modem offers speeds in multiples of T1/E1.
- For an optimal connection, the local loop of a DSL must be less than 3.5 km (2.2 miles).
- Cisco has modified HDLC and made a proprietary protocol based on it.
- WAN Addressing is based on logical addresses for IP networks.

ON THE CD

All code listings, figures, and tables presented in this book can be found on the book's companion CD-ROM.

2 Design Considerations

IN THIS CHAPTER

- Enterprise-wide Networks
- Protocol Media Transport Problems
- PDIOO
- Network Needs Analysis

The first step towards creating a robust design is to understand your customer requirements. The network design should be such that it fulfills the customer's immediate requirements and maps them to the long-term business goals. Before you begin the actual design, collate data about the customer based on certain parameters such as existing network infrastructure, nature of business operations, and corporate policies.

ENTERPRISE-WIDE NETWORKS

An enterprise-wide network connects the Local Area Networks (LANs) of each office in an organization, spanning different geographical locations. These offices may be located in close proximity to each other or in different cities and countries across the globe. An enterprise-wide network can also be called an internetwork.

Based on the existing infrastructure, enterprise-wide networks can be classified into small, medium, and large. It is important to remember that the perception of small, medium, and large networks may vary with respect to different factors, which could include the level of technological advancement of a country and the nature of operations of an organization. For example, for a technologically advanced country, a network with 30 locations would be considered small. For a less technologically developed country, the same network would be considered large.

The nature of operations can have an impact on the size of the network. For example, a complex network for a defense department may have only 30–40 locations while a media company may have 60–70 locations. Given the complexity of operations of the defense department, the network may be much larger than that of the media company.

Small-sized Networks

Small-sized networks cater to a small number of users and are confined to a building or a cluster of buildings in close proximity. In the context of a LAN, a network of 30 to 40 users is considered small. Similarly, a network of 100–150 users within one or multiple locations or premises in close proximity can also be called a small-sized network. The main advantage of using small-sized networks is easy manageability because they are less complex. Figure 2.1 depicts a small-sized network.

In Figure 2.1, there is one central server and three managed stackable hubs. Each of the stackable hubs is connected to individual workgroups, Workgroup 1, Workgroup 2, and Workgroup 3. There is also a print server connected to the network. These elements together form a small enterprise network.

Let us take an example of F&G Inc., a pharmaceutical company. In addition to the headquarters at Connecticut, there are two branch offices in Massachusetts and Georgia.

Examples of devices associated with small-sized networks are stackable managed hubs, file servers, print servers, mail servers, DHCP servers, DNS servers, and proxy servers.

NOTE

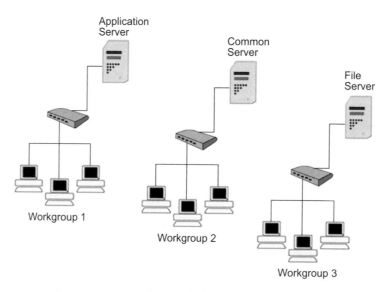

FIGURE 2.1   A small-sized network showing a server and three hubs.

Medium-sized Networks

A medium-sized network caters to approximately 1000 users, spread across 20–25 locations, with each location being connected to the headquarters. Figure 2.2 depicts a medium-sized network.

Figure 2.2 depicts two locations; the building in the first location has three floors, and the second building has two floors. The two locations are connected through a router, which is connected to the Internet. The floor of each building has a managed stackable switch that connects various network devices such as workstations, and file, Web, and print servers.

Take the example of F&G Inc., which has expanded its scale of operations. The organization now has 900 employees in 18 offices across the country. The 15 sales offices are connected to the headquarters in Connecticut through a 64 Kbps leased line. The manufacturing units in Massachusetts and Georgia are connected via a 256 Kbps leased circuit.

The headquarters at Connecticut has a LAN consisting of a 10BaseT hub. There are seven Ethernet segments at about 45 percent utilization. All seven segments are connected with two Cisco 3600 routers.

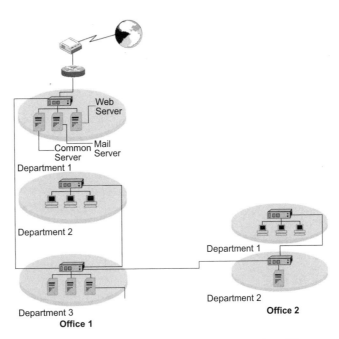

FIGURE 2.2   A medium-sized network connecting different departments.

Large-sized Networks

A large-sized network is a multilocation network with more than 1000 users in more than 50 locations, often spanning many countries. Examples of organizations using these networks are government organizations and multinational corporations. These networks operate to provide services to customers, employees, and vendors. Figure 2.3 depicts a large-sized network spread over many locations.

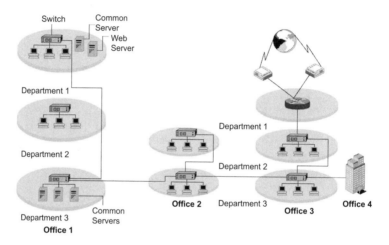

FIGURE 2.3   A large-sized network connecting departments in different locations.

Connectivity within floors in a building is established with stackable managed switches, and the inter-office connectivity is established through multiport Gigabit switches.

One of the locations has a remote router, which is connected to two leased line modems. These modems provide fast connectivity to all the offices via the leased lines.

PROTOCOL MEDIA TRANSPORT PROBLEMS

Every network has some typical problems associated with it, irrespective of its size. To reduce the complexity associated with customer problem identification, analysis, and solution design, Cisco has classified network problems into three categories: media, protocols, and transport. Figure 2.4 depicts the framework in the form of a triangle.

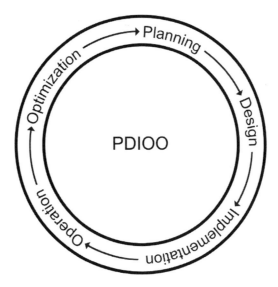

FIGURE 2.5   PDIOO network lifecycle showing interrelated stages.

PDIOO

Planning, Design, Implementation, Operation, and Optimization (PDIOO) methodology represents a network lifecycle—from the time of planning until the time the network is ready for redesign. The five distinct but interrelated stages in the PDIOO lifecycle are depicted in Figure 2.5.

This lifecycle is applicable for any type of network: small, medium, or large.

Planning

Planning is the first step in the PDIOO methodology. This is also the most important phase because it allows you to assess the current network setup. The new setup would be implemented as planned only if the existing systems are compatible with the upgraded number of applications. Planning sets the course of the design, and any flaw at this stage is visible only when the network is operational. You should let the customer test the design before it is put into operation. The User Acceptance Test ensures that the design achieves the business targets.

Design

Based on the information gathered in the planning phase, design a blueprint of the proposed network using graphs, flowcharts, and models. The blueprint will contain the minutest details pertaining to the proposed network. It will include descriptions such as the types and numbers of devices and their placement and the layout of the cables. To present your design, use a sample demonstration or simulation of the final network model. A good design model serves the basis for the implementation stage of the network designing methodology.

Implementation

When the customer approves the design, you need to implement the design model. The implementation step is important to the PDIOO process because this step ensures that the designed plan is functional. After implementing the design and giving it a test run, you can get an idea of the possible errors and omissions in the design. During this stage, you should build a pilot of the actual network model. This pilot is then given a test run to check for any flaws, loopholes, and bugs. The errors detected in the pilot are a good indication of the weaknesses in the network design model. Detecting these errors in the test run will give time to make changes in the design model before the network is completely operational.

Operation

The operation stage of the PDIOO methodology is the phase when the operational network design is launched. The performance levels of the network should be monitored 24 x 7, and any flaws detected in this phase should be rectified. This may require some changes in the network design. The information gathered in this phase provides an input for the next step of the network design methodology: optimization.

Optimization

During this phase of the PDIOO process, use different network management tools to optimize network performance. It is at this stage that you detect serious network problems pertaining to media, transport, and protocols, and possibly opt for a network redesign or upgrade.

NETWORK NEEDS ANALYSIS

When you design a network, ensure that the design fulfils the customer's immediate requirements and matches the increase in the scale of business operations. In other words, whether you design a new network or revamp an existing one, the output should reflect an understanding of the organization's current and long-term business goals.

As a network designer, you have two primary objectives while analyzing your customer's needs:

■   Evaluate the current corporate profile. This is related to collecting both nontechnical and technical information pertaining to the customer's business goals, issues, and constraints.

■ Evaluate the future requirements, design needs, and expectations.

Start with the actual design process only after collecting the aforementioned information and documenting it. This will enable you to create a solid and viable network design model that will satisfy the customer's requirements related to functionality, performance, security, and scalability.

Current Assessment

Assessment of the current network involves a thorough investigation of the customer's inventory and network performance. This serves as a reference point to start designing the network blueprint.

Begin with the actual design process only after collecting and documenting all data pertaining to the customer's needs. You need to understand the parameters and assess your customer's current and future requirements accordingly.

The first phase of analyzing your customer's requirements involves collecting data about the existing infrastructure under two broad categories:

Technical data: Refers to all network-related information such as protocols in use, broadcast rates, current network traffic, collision rates, and network bottlenecks.

Administrative data: Refers to all information pertaining to the customer's business type, current corporate and geographical structure, business goals, and other business issues such as organizational policies that might affect your design.

Figure 2.6 depicts different parameters for collecting technical and administrative data.

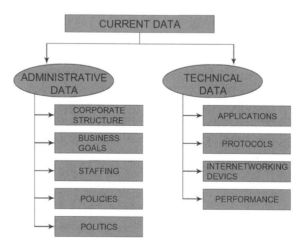

FIGURE 2.6 Parameters for gathering administrative and technical data for a customer.

When you have collected the administrative and technical information pertaining to the current network, document the existing network on the basis of certain parameters like network topology, addressing schemes, and traffic flow.

The network of an ISP is functionally different from any other enterprise. As a result, the assessment of the current infrastructure of an ISP is different from other enterprises. You need to look at the current assessment of each in detail.

Collecting Administrative Data

Before developing a design solution for a network, assess your customer's current business type, business goals, and constraints. Analyzing your customer's administrative structure, needs, and goals form the core for creating an effective design plan because the information gathered from the analysis will influence the design of the network. For example, when you know that the customer expects to increase the scale of operations by 50 percent each year and might enter an alliance with other partners, suggest a modular network design to accommodate future expansion.

Perform a detailed administrative analysis to collect data related to the existing organizations and proposed organizational structure, policies, procedures, and practices. In addition, you should collect data that reflects the customer's corporate mission, geographical structure, and current and future manpower; all of that can influence your internetwork design. Various types of administrative data that you need to accumulate before starting the design are discussed in this section.

Nature of Business and Long-term Goals of the Organization

With knowledge about your customer's business and external relations, suggest an optimal solution for business operations.

To determine the type of business your customer is engaged in and the company's estimated business goals and growth expectations, use the parameters listed in Table 2.2.

TABLE 2.2   Parameters for Determining Business Goals

Customer Projections	Projection Parameters
Scope of customer's business	• Growth projections for the next five years • Customer's industry standards, clients, partners, and potential competitors in the field • Customer's performance in the market
Business processes	• Roles and responsibilities of various employees such as data administrators, data processors, and data security administrators • Business policies followed in the organization
Application requirements	• Applications required to execute business operations • Network resources required by the users • Type of security applications

Business Structure

Analyzing business structure is necessary because each level of the corporate structure has different network requirements. You will recognize the management hierarchy by understanding the corporate structure. For example, if you have to design a network for a news production company, you might have different design considerations for the production department as compared to the accounts or HR department. This is because the production department would have a requirement of a high-end network with intelligent hardware and software having a high bandwidth and Quality of Service (QoS). This might be different than the requirement of the HR department because it would not require such high-end network service. In your first meeting with your customers, ask them to explain the organizational structure of the company. Analyze the hierarchical corporate structures and identify relationships within corporate entities. It is a good idea to gain an understanding of how the company is structured in departments, vendors, partners, and field or remote offices because your final internetwork design should reflect the corporate structure. You will locate the user communities and characterize traffic flow by understanding the corporate structure.

Current and Future Staffing

Your network design model should be scalable so that it accommodates an increased number of users. When the current and future staffing requirements are available, you can project the company's vision and design a network to match their goals.

To determine the company's current staffing levels and their future expectations, ascertain the following information:

- Current staffing level
- Number of technical personnel for internetworking within the organization
- Expected changes in the job functions of staff with the introduction of the proposed network
- Expected expansion or elimination of staff
- The identity of the IT or business executive who will assist you with the design

Business Policies and Politics

Gather information regarding past successes and failures because this will aid you in assessing the bottlenecks that might come up in the design. You can ask various questions of your customer:

- Are the employees of the company open to implementing the changes?
- What is the current level of in-house expertise pertaining to networks?

Geographical Structure

The proposed design of the network should be such that it can accommodate the organization's growth. Further, it must accommodate disaster recovery and load sharing between branch offices, if the customer plans to have them. For this, collect information related to the location of the client's user communities and map this information to the growing business structure of the organization.

Collecting Technical Data

After gathering the required administrative data from the customer, carry out a comprehensive technical assessment of the current network infrastructure and future requirements. This will help you to identify the customer's IT infrastructure such as hardware, network circuits, internetworking devices, processes, applications, and facilities of the computing (LAN/WAN) environment. This will help you detect any network bottlenecks and issues that might arise while implementing the proposed design. A combination of administrative and technical data will help you design a network that will serve the needs of the customer.

After knowing the business your customer is engaged in, assess the type of applications in use. For example, an airline reservation department of XYZ Airlines has a central database accessible from all the regional centers of the country. This means that applications, made by such companies as Oracle, Ingress, or Sybase would be running on the network. Similarly, if the customer is a service provider, analyze the type of LAN/WAN connections, bandwidth, and devices that should be used. Use this information to identify how the applications and the traffic flow can be affected by the bandwidth.

Collect technical data from your customer with respect to the following aspects:

- Current applications
- Network performance
- Network management
- Network security

Current Applications

Identify the customer's current applications and requirement for future applications. You can ask questions such as:

- What are the types of applications used on the network—distributed, collaborative, or centralized?
- Is it a computing application (as opposed to e-mail, voice applications, video streaming, and so on)?
- Is the current application a customized or standard application?

- How many workstations are required?
- Are there any flaws in the application?
- What is the response time for traffic like AppleTalk, DHCP, NetIOS, TCP/IP, NetWare client traffic?
- Is the application mission critical?
- How many network devices are in use on the network?
- How do different applications and traffic levels affect the flow of information in the company?
- What is the percentage of delay-sensitive traffic, whether or not the customer is already using any QoS features?

After collecting this information, determine the LAN and WAN bandwidth requirements of the proposed network, the protocols that run on the network, and the density of traffic that the new network needs to be able to handle. For example, networks with linear traffic, such as file transfer and e-mail, would be different from the network simultaneously carrying converged, bandwidth-intensive, and delay-sensitive traffic such as voice, video, and data. To ensure that they run properly, allocate enough bandwidth to meet all requirements.

Network Performance

Collecting data pertaining to network performance provides a detailed view of the overall performance and application use of the network. With this information, you can create a baseline model of network activity and develop a topology of the customer's current networking environment. In addition, it will help identify the possible network capacity issues or shortcomings that you might confront in the future.

To collect this information, deploy tools such as sniffers to gather information on the network across all data elements and devices. Do the following to assess your customer's network performance:

1. Ask for baseline reports.
2. Measure key network performance indicators such as network utilization, latency, and errors.
3. Identify performance bottlenecks and intermittent problems that slow the network response time.
4. Measure the traffic flow within segments and monitor for errors during peak loads at different times of the day and week.
5. Detect changes when new applications are deployed or when the network expands.

To collect data for Net audits, use software such as The Network Data Collector, a telnet- and SNMP-based data collector for Cisco devices.

NOTE

Network Management

Collecting information pertaining to network management provides a detailed description about effectively managing your customer's networks and nodes. In addition, you can validate the health of existing technologies and practices and minimize the threats and risks with an ever-changing infrastructure. You can query the customer with respect to these subjects:

- Type of network management procedure/method/tool to be implemented
- Technical expertise of the customer's IT department
- Technologies, tools, or software being used to assess network management tasks
- Future growth patterns and customer adoption rates of new infrastructure technologies

Network Security

With the increasing number of threats to network security, it is important to document the network security policies, measures, and devices used by your customer. By collecting data about network security initiatives, you will understand the security challenges faced by the customer and the security features required in the proposed network design.

You need to identify the weaknesses and deficiencies in a given system. Talk with the design, security, and operations teams to collect information related to current network topology, security policies and requirements, and traffic profiles for existing and planned network services. You can ask these questions:

- Are the network equipment and cables physically protected from security breaches?
- What are the network security threats and security risks that are a priority to the customer?
- How many attempts have there been to breach network security, and what types of attempts were they?
- How much traffic is local to the segment and how much traffic is external?

NOTE

You can use vulnerability assessment scanners such as Cisco's secure scanner, Stat Scanner, ISS Internet Scanner, and SAINT Vulnerability Scanner.

You also need to evaluate each application's architecture, design, and function; the platform it runs on, the networking services it uses, and any operating platforms services used. By gathering this information, you can assess the weaknesses in your application, how they are caused, the risk levels, and the steps that can reduce them.

Diagramming the Network

After collecting the administrative and technical information pertaining to the customer's current network, portray a network on the basis of certain parameters. Cisco provides 12 steps to document a customer's existing network, which are:

- Describe customer's existing network applications
- Describe existing network protocols
- Document network topology and addressing schemes
- Identify business constraints
- Characterize network performance
- Document network availability
- Document network reliability
- Document network utilization
- Document the status of existing routers
- Summarize the existing network infrastructure
- Determine potential bottlenecks
- Document the health of the existing network

Customer's Existing Network Applications

Collecting data on existing network applications is of utmost importance because the network design has to be built based on the data. Every standard application has well-defined benchmarks in terms of response time and other factors. By evaluating the current applications, you can determine if:

- The current network is adequate or if there is a need for additional network equipment
- The existing network would support more applications and users or new additions are required

Create a table like Table 2.3 to document the customer's applications.

TABLE 2.3 A Form for Documenting Existing Network Applications

Application Name	Application Type	Number Users	Number of of Hosts or Servers	Segment	Remarks

Existing Network Protocols

Network protocols are important for a network from the point of view of scalability and growth. For example, IPX/SPX cannot scale because it contains a flat addressing scheme devoid of any hierarchy as compared to TCP/IP. IPX/SPX uses Service Advertising Protocol (SAP). SAP works by sending broadcasts that "advertise" the services available on the server, and because of the increased traffic, is unsuccessful in WAN environments. While documenting the network protocols, document both the routing and routed protocols. Create a table like Table 2.4 to document existing network protocols.

TABLE 2.4   A Form for Documenting Existing Network Protocols

Protocol Name	Protocol Type	Number Users	Number of Hosts or Servers	Remarks

Network Topology and Addressing Schemes

Collecting data about the network topology and addressing schemes is important as it helps determine:

- The number of users the network segment can accommodate
- The class of the addressing scheme
- Whether *Variable Length Subnet Masks* (VLSMs) are present
- The network topology—point-to-point, point-to-multipoint, full-mesh or partial-mesh
- Type of computing—centralized, decentralized, or collaborative
- Number of office locations

To document the existing network topology, ascertain the routes between gateways, networks, and hosts in the customer's network. For this, you have to assess:

Existing network topology: Refers to the type of segments in a customer network, segment speeds, naming schemes of routers and switches, and names and addresses of all internetworking devices and servers

Addressing scheme: Includes documentation of IP addressing schemes and models such as network addresses and subnet masks for IP

Other network-related information: Includes documentation about information pertaining to issues such as traffic flow and architecture

A network topology map should include:

- An illustration of the logical architecture of the network (for example, bus, star, ring, mesh, partial mesh)
- Data-link layer technology for WANs and LANs (for example, frame relay, ISDN, 10 Mbps Ethernet)
- Locations of routers and switches
- Locations and accessibility of the nodes in VPNs that connect corporate sites using a service provider's WAN
- Locations and reach of the nodes in Virtual LANs (VLANs)
- Topology of firewall security systems
- Location of any dial-in and dial-out systems
- Location of enterprise servers and users
- Redundant devices and links

Business Constraints

When identifying business constraints, review the administrative data that you have collected. It is essential to understand the factors that have influenced the current network design and then design a network that achieves the desired scalability. Document business constraints such as organizational policies, budgets, staffing, and timelines. You have to assess the:

- Corporate structure of your customer and extract the business profile information
- Mission-critical operations
- Policies that affect the implementation of the new design
- Technical know-how of the customer

After gathering this information, you will be able to identify the issues and challenges of your customer. This will facilitate the design acceptance and approval. In addition, you will know more about the decision makers in the organization. This information will aid in meeting customer requirements within the budget.

Network Performance

Collecting information about network performance would enable better capacity planning. Information gathered by network and protocol analyzers will help achieve a better design with a higher QoS. You can also identify:

- Bottlenecks and congestion points the network is facing
- Problematic domains

To document details about network performance, measure the response time between hosts, devices, and applications running on the network. To characterize the performance of a network, organize the data as shown in Table 2.5.

TABLE 2.5   A Form for Documenting Details About Network Performance

Host A	Host B	Host C	Host D
Host A			
Host B			
Host C			
Host D			

Network Availability

Collecting data about network availability is important for every mission-critical network. Determine if the company has back-up network links, if there are any redundant network links, and what the Mean Time Between Failures (MTBF) is and how can it be minimized. To document the network availability, gather information pertaining to network downtime and MTBF for the enterprise network. You can collect data pertaining to the following factors:

- Critical segments of the network
- Causes of failure of those segments
- Duration of network failures
- Hourly cost by department for any network breakdown
- Hourly cost to the company for such network breakdown

To characterize network availability, organize the data as shown in Table 2.6.

TABLE 2.6   A Form for Documenting Details About Network Availability

Segment	MTBF	Date of Last Downtime	Duration of Last Downtime	Cause of Last Downtime
Internetwork				
Segment 1				
Segment 2				
Segment 3				

Network Reliability

To measure network reliability, document the network traffic patterns such as peak utilization, average network traffic, number of broadcast and multicast frames, and sampling rate. By measuring the network reliability, you can ascertain potential problems with the existing network. Network reliability measures the transmission errors that occur when the network is functional. You can measure network reliability through network management tools such as protocol and network analyzers that are deployed within LAN and WAN segments to examine the network traffic cycles. Document the values as listed:

- Total amount of data transferred in megabytes
- Number of frames passed during data transfer in a network
- Number of Cycle Redundancy Check (CRC) errors
- Number of MAC layer errors (collisions, Token Ring soft errors, and FDDI ring operations)
- Number of broadcast/multicast frames

To characterize the network's reliability, organize the data as shown in Table 2.7.

TABLE 2.7 A Form for Documenting Details About Network Reliability

Network Segment	Average Network Utilization	Peak Network Utilization	Average Frame Size	CRC Error Rate	Mac Layer Error Rate	Broadcast/ Multicast Rate
Segment 1						
Segment 2						
Segment 3						

Use the formulas as listed to calculate the values in Table 2.7.

$$\text{Average Network Utilization:} \frac{\text{Hourly average utilization}}{\text{Number of hourly average}}$$

Average Network Utilization = Sum of hourly average utilizations / Number of hourly averages

For example, if the hourly average utilizations for 3 hours is 10 Mbps, 15 Mbps, and 5 Mbps for each hour respectively, Average Network Utilization =(10+15+5)/3 = 10 Mbps Peak Network Utilization: Record the highest hourly average

$$\text{Average Frame Size:} \frac{\text{Total number of mega bytes of data transfer on the network}}{\text{Total number of frames}}$$

$$\text{CRC Error Rate: } \frac{\text{Total number of CRC errors}}{\text{Total number of frames sent}}$$

$$\text{MAC Layer Error Rate: } \frac{\text{Total number of MAC layer errors}}{\text{Total number of frames}}$$

$$\text{Broadcast/Multicast Rate: } \frac{\text{Total number of broadcast multicasts}}{\text{Total number of frames}}$$

Network Utilization

To effectively utilize a network, collect information pertaining to the current and maximum bandwidth, current and minimum latency, bottlenecks, and the extent of congestion in the network. For this, use network monitors or network measurement tools.

To characterize the network utilization, organize the data as shown in Table 2.8.

TABLE 2.8 A Form for Documenting Details About Network Utilization

Protocol	Relative Network Utilization	Absolute Network Utilization	Average Frame Size	Broadcasts/ Multicasts Rate
IP				
IPX				
AppleTalk				
NetBIOS				
SNA				
Other				

To calculate the values listed in Table 2.8, use the following formulas:

Relative Network Utilization: First, calculate the percentage of each type of protocol traffic on a segment. Then, perform the following calculation:

$$\frac{\text{Utilization amount for a segment from a specific protocol}}{\text{Total bandwidth use for a segment}}$$

$$\text{Absolute Network Utilization: } \frac{\text{Bandwidth use of each segment}}{\text{Size of the actual bandwith}}$$

Let us say the Actual Bandwidth of a segment = 500 Mbps and Total bandwidth being used on the segment = 200 Mbps, and the bandwidth corresponding Protocol A on the segment = 5 Mbps. Then, the Relative Network Utilization for that protocol on that segment is 5/200 = 2.5% and the Absolute Network Utilization for that protocol = 5/500 = 1%.

Status of Existing Routers

To characterize a network, collect data about the status of the existing routers and traffic. This will enable you to consider the scalability of the existing routers in a new network. You can also determine the number of available ports for future growth and the number of modules to be added if it is a modular router. With this information, move routers to other locations if they are not scalable and use high-end routers at a central or strategic location.

You can check the status of major routers by using SNMP or some basic show commands created by Cisco IOS software. These commands are listed in Table 2.9.

TABLE 2.9   Basic show Commands

Command	Used For
Show interfaces	Displays the type, identity, and operational status of all interfaces attached to a device
Show processes	Displays active processes and CPU utilization for 5-second, 1-minute, and 5-minute intervals
Show buffers	Displays information to identify over-utilization problems in main memory

To check the status of existing routers, organize the data as shown in Table 2.10.

TABLE 2.10   A Form for Documenting Details About Existing Routers

Router Name	5-Minute CPU Utilization	Output Queue Drops Per Hour	Input Queue Drops Per Hour	Missed Packets Per Hour	Ignored Packets Per Hour	Remarks
Router 1						
Router 2						
Router 3						

Assessing Network Management Tools and Systems

By understanding the customer's current management system, you can analyze the expectations for the new network management system. Document the types of network management tools and systems that are in use as well as the tools to assess the quality of the network devices. In addition, list the platforms for these tools to run on and other maps or reports in use. Current industry standards allow you to adopt a single solution such as HP OpenView® or CiscoWorks 2000 as a network management strategy.

CiscoWorks is a family of comprehensive network management tools that allows you to assess and manage the advanced capabilities of the Cisco AVVID architecture through a single console. The tools provide innovative ways to consistently centrally manage critical network characteristics, such as availability, response, flexibility, and security.

Potential Bottlenecks

Bottlenecks in the network occur mostly due to undesirable traffic types, broadcasts and multicasts, bad applications, and bugs. After gathering information about the current network, you would be able to design a congestion-free and foolproof network. In addition, it helps you to estimate the number of applications and users that the new network will be able to accommodate in the future and their impact on the network.

You can stress test the network with packet generator tools and protocol analyzers to find out how the network would function in different traffic conditions. Use network analysis probes to provide detailed information on network dataflow, capacity, downtime, and other snags. After getting this information, you can detect the problem areas, identify their causes, and recommend solutions to counter them.

You can use protocol analyzers or sniffers to determine the amount of local traffic. Sniffers also determine the amount of traffic that is not local, that travels to other network segments, and that comes from other network segments. On TCP/IP networks, these are known as packet sniffers.

Health of the Existing Network

When you have gathered information about the current network, determine the health of the existing network. Ask for network health reports from your customer. This report forms a baseline that serves as a guideline on how the network would behave at peak and off-peak traffic conditions. This will help you design a network with better capacity planning and adequate QoS and bandwidth management.

make sure to include all necessary functionality in the network design. For example, if new technologies such as videoconferencing or LAN telephony will be needed by the company, it will be much less expensive to include them in the original network design than to add them after the network has already been built.

As a result, identifying manageability requirements of the customer is important. As the network grows, it becomes more complex in nature as it has to accommodate both legacy and future environments. After you have assessed the client's manageability requirements using various network management tools, document the details for later reference.

You can identify the customer's manageability requirements using the factors listed in Table 2.12.

TABLE 2.12   Factors for Identifying Manageability Requirements

Identifying Requirements Related to:	Tasks Involved:
Performance management	Monitoring the performance of various devices in a network, which helps while planning for future expansion
Configuration management	Identifying and setting up configuration items in a network, recording and reporting the status of configuration items and requests for change, and verifying the completeness and correctness of configuration items
Fault management	Identifying, solving, and reporting errors and problems
Security management	Monitoring and testing security policies, distributing passwords, managing encryption and decryption keys, and auditing adherence to security policies
Accounting management	Accounting of network use to allocate costs to network users and/or plan for changes in capacity requirements

Identifying Application Requirements

If you have collated the application requirements of your customer, specify throughput and data rates required for the proposed network. Set the application requirements of the network that you have to design. Applications have a huge impact of the rate of data flow, and as a result, you need to determine how the new design will affect the flow of data throughout the facility. To determine the application requirements of the new network, find:

- Application names and types
- Protocol names and types
- Constraints of current protocols
- Appropriate networking technologies to be implemented
- Number of end users who will use the application and protocols
- Peak hours of usage of new applications
- Traffic characteristics of the applications
- Delay-sensitivity of the applications

Identifying Network Traffic Requirements

Identifying network traffic requirements means determining network traffic load and traffic behavior. One of the most difficult aspects of determining a customer's requirements is to analyze how the applications and the protocols behave.

To identify network traffic patterns, characterize:

- Traffic load:

  - Minimum and maximum traffic loads
  - Effective routing procedures
  - Network bandwidth

- Traffic behavior:

  - Broadcast/multicast behavior
  - Supported frame sizes
  - Windowing and flow control
  - Error recovery mechanisms

To characterize the network traffic, use various network analyzers such as Cisco's NETSYS Baseliner tools, Cisco's NetFlow, and CiscoWorks.

Identifying Network Performance Requirements

To identify network performance requirements, deploy various network management applications that monitor and measure performance of the network across LAN and WAN links, and also the services running on that network. You can determine network performance in various areas, as follows:

- Response time
- Accuracy

- Availability
- Maximum network utilization
- Throughput and efficiency
- Latency

Today, baselining and capacity planning for the network resources have become a critical part of systems integration requirements. The key focus is to identify utilization levels of resources (disk space, system memory, and network bandwidth) at the time of integration and assess them regularly. This ensures that the new design accommodates increased utilization and shows enhanced performance.

Identifying Network Security Requirements

To identify the security requirements of the new network, carry out the tasks as listed:

- Assess the level and type of security required for the network.
- Identify the possible security risks.
- Determine the requirements for outsiders to access data from the network.
- Determine authentication and authorization requirements to access the network.
- Identify the requirements for authenticating routing updates received from access routers.
- Identify requirements for host security such as physical security of hosts, user accounts, and access rights on data.

SUMMARY

In this chapter, we learned about the enterprise-wide networks and their classifications. We also reminded you that different networks have different requirements and use different devices to operate. In the next chapter, we move on to different models of network topology.

POINTS TO REMEMBER

- Enterprise-wide networks can be classified into small, medium, and large.
- Small-sized networks are easy to manage, less complex, and use devices such as stackable managed hubs, file servers, and print servers.
- A medium-sized network encompasses around 1000 users spread across 20–25 locations, with each location being connected to the headquarters.

- A medium-sized network uses devices such as workstations, and file, Web, and print servers.
- A large-sized network is a multilocation network with more than 1000 users in more than 50 locations, often spanning many countries. It uses stackable managed switches, multiport Gigabit switches, routers, and leased-line modems.
- Different types of media available for connecting devices are twisted-pair, coaxial, and fiber-optic cables.
- Media problems occur when too many devices contend for access to a particular LAN segment or when the length of the cable exceeds the specified limit.
- The problems include increased number of collisions on Ethernet networks and long queues for the token on a Token Ring or FDDI network.
- A solution to media problems is the use of LAN switching to create separate collision domains.
- Protocol problems can lead to network congestion and can be solved by dividing a network into separate segments using one or more routers.
- The five stages in PDIOO methodology are planning, design, implementation, operation, and optimization.
- A design begins after assessment of a customer's needs. The first phase involves collecting administrative and technical data.
- The nature of business and long-term goals of the business should be known to the network designer.
- A network design model should be scalable, so that it can accommodate an increase in the number of users in the future.
- Collect technical data from your customer with respect to current applications, network performance, network management, and network security.
- Diagramming of the network is done on the basis of 12 factors provided by Cisco. These are the customer's existing network applications, existing network protocols, network topology and addressing schemes, business constraints, network performance, network availability, network reliability, network utilization, status of existing routers, existing network infrastructure, potential bottlenecks, and health of the existing network.

3 Network Architecture Models

While designing networks, the first task is to collect and analyze customer's current needs and future requirements. After a thorough study of the customer's requirements is performed, the next task is to select the appropriate network topology. There are three different types of network topology models: the *hierarchal*, *fault-tolerant*, and *secure* models. An organization should select an appropriate topology model to help optimize various network concerns such as scalability, efficiency, adaptability, security, and dependability.

HIERARCHICAL MODEL

A hierarchy helps you to summarize a complex collection of details in an understandable structure. All of us have encountered the meaning of hierarchy in our lives—at home or work. The hierarchical system at work directs you to follow the instructions of your boss or give instructions to your subordinates.

For example, any larger organization has a CEO at the top of the hierarchy. The Sales, Operations, and HR managers function at the second level of the hierarchy. These managers delegate responsibility to their individual teams. This results in smooth functioning of the entire organization. As a result, the organization has the CEO as the strategic head, the Sales, Operations, and HR managers to organize workflow, and the remaining employees involved in day-to-day operations of their respective departments.

The three layers of the Cisco hierarchical model are *core*, *distribution*, and *access*. The core layer is the topological center of the network and provides high-speed switching, redundancy, high availability, and bandwidth to traffic passing through it. It does not process or route any packets. The distribution layer is responsible for packet filtering and routing. The core layer provides high density and processing capacity-per-interface for packet manipulation. The access layer is the outermost layer and enables end users of the network to connect to each other.

This segmentation of the network into layers is logical, not physical. Each layer performs a specific set of tasks. The functions of these layers may overlap during a physical implementation of a hierarchical network.

The hierarchical model has been designed on the premise that the volume and patterns of network traffic impact network performance. As a result, the three layers have been designed to efficiently manage network traffic, reduce the network load, and increase scalability.

A hierarchical model is designed to enable high-speed data transmission. For example, a financial organization that provides speedy money transfers to its customers opts for the hierarchical model. If you need to wait a half an hour to complete an online transaction with your bank, you will consider that poor service. Therefore, banks and other financial organizations adopt the hierarchical model.

When designing and implementing the network, ensure that the layers are placed such that the traffic is efficiently forwarded between layers. This is because the network rejects any unnecessary packet that a lower layer forwards to an upper layer; that decreases the upload time. Figure 3.1 depicts the core, distributed, and access layers of the Cisco hierarchical model.

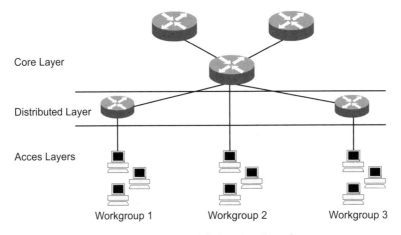

FIGURE 3.1   Cisco hierarchical model showing three layers.

The Core Layer

The core or backbone layer is at the top of the hierarchical model and forms the crux of the entire network. This layer is a high-speed switching backbone that transports large volumes of traffic reliably and quickly. It provides wide area connections and moves large payloads between geographically distant sites. The traffic transported across the core is between end users and enterprise services. The links in the core layer are point-to-point.

Examples of enterprise services are the Internet and e-mail.

NOTE

The core layer is responsible only for providing high-speed transport. As a result, there is no room for latency and complex routing decisions pertaining to filters and access lists. These are implemented at the distribution layer. Therefore, protocols with fast convergence times are implemented at this layer. Quality of Service (QoS) may be implemented at this layer to ensure higher priority to traffic that may otherwise be lost or delayed in congestion.

The core layer should have a high degree of redundancy. This is because if there is a breakdown in the core layer, it will adversely affect the entire network. For example, referring to Figure 3.1, if the core layer router fails, the routers in the distribution layer and the workgroups in the access layer will be affected. If redundancy is incorporated, an alternate path is used to transport data in case of any failure in the active path.

The performance of the core layer also depends on the diameter of the network design. The number of hops from one edge router to another edge router is called the *diameter*. The diameter within the hierarchical network should remain consistent for good performance of the core layer. This means that the distance from any end station to another station should be equal. Any changes in the diameter would result in irregular movement of data, which results in problems such as over-utilization of a few nodes, frame delays, and data loss. Therefore, the overall performance of the core and hierarchical layers deteriorates. The consistency of the diameter in the core layer is depicted in Figure 3.2.

The core layer implements the following functions:

- High-speed switching of large volumes of traffic
- Redundancy and fault tolerance

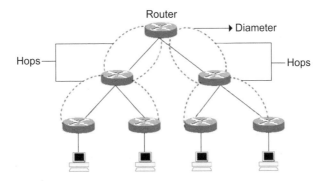

FIGURE 3.2   Consistency of the diameter in the core layer.

- High network availability and reliability
- Load sharing
- Rapid convergence
- High bandwidth availability
- Low latency

Table 3.1 depicts the do's and don'ts to be kept in mind when implementing the core layer.

TABLE 3.1 Do's and Don'ts of Implementing the Core Layer

Do	Don't
Use Fast Ethernet, Asynchronous Transfer Mode (ATM), and Fiber Distributed Data Interface (FDDI)	Use Access Control Lists (ACLs)
Use routing protocols with lower convergence time	Implement Virtual Local Area Network (VLAN) routing and packet filtering
Use fast and redundant data link connectivity	Expand the layer
Maintain consistent diameter	Implement work group access

The Distribution Layer

The distribution or *policy* layer is placed between the access and core layers. This layer implements network policies and controls network traffic and data movement. In addition, it performs complex, CPU-intensive calculations pertaining to routing, filtering, inter-VLAN routing, ACLs, address or area-aggregation, and security, and identifies alternate paths to access the core. The distribution layer also determines the best possible path to forward end-user requests to the core layer. The core layer then transports end-user requests and responses to and from enterprise services.

Members of this layer include most local servers on the LAN, routers, and backbone switches. Various features are associated with the distribution layer, as follows:

- Access lists and QoS
- Packet filtering and queuing
- Redistribution of routing protocols
- Inter-VLAN routing
- Broadcast and multicast domain definition
- Address or area aggregation
- Departmental or workgroup access
- Media transitions

- Redistribution between routing protocols
- Load balancing
- Router summarization and route filtering
- Security and network policies
- Address translation and firewalls
- Controlled access to core layer resources

The Access Layer

The access or workgroup layer is the outermost layer of the Cisco hierarchical model. On this layer, all end users are connected to the LAN. The end-user workstations and local resources such as printers are placed at this layer. Routers serve as gatekeepers at the entry and exit to this layer and ensure that the local server traffic is not forwarded to the wider network. The access layer controls user and workgroup access to internetwork resources. Other functions performed on this layer are sharing and switching of bandwidth, MAC-layer filtering, and micro segmentation.

The features associated with the access layer are:

- Access control lists
- Traffic filtering
- Segmentation
- Workgroup connectivity with the distribution layer
- Creation of separate collision domains
- Static routing
- MAC-layer filtering
- Switching and sharing of bandwidth
- Dial-on-Demand Routing (DDR) and Ethernet switching

Table 3.2 shows the functions of the three layers of the hierarchical model.

TABLE 3.2 Layers and Their Functions

Layer	Functions
Core layer	• High-speed switching • Fault tolerance • Maintaining consistent diameter
Distribution layer	• Policy-based connectivity • Area aggregation • Load balancing
Access layer	• Workgroup access • Segmentation • MAC-layer filtering

Figure 3.3 depicts the three layers of the hierarchical model along with their key functions.

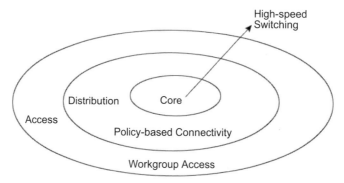

FIGURE 3.3   The hierarchical model showing the three layers.

Benefits of the Hierarchical Model

An ideal network should provide optimal bandwidth utilization, efficient management of network traffic, and maximum uptime. Keeping these factors in mind, Cisco introduced the hierarchical design model. This model enables you to design a scalable, reliable, and cost-effective hierarchical internetwork. The benefits of a hierarchical model of network topology are in the following areas:

- Manageability
- Performance
- Cost
- Scalability
- Speed

Manageability

Networks based on the hierarchical model are easy to manage, troubleshoot, and support. The functions of the three layers are exclusive to their individual layer, and as a result, it is easy to determine and isolate problems within each layer. If there is an error in the functioning of a workstation or network device at any of the three layers, it is easy to diagnose and troubleshoot without disturbing the other layers. For example, if a workstation placed at the access layer fails, you can easily troubleshoot without disturbing other layers. Figure 3.4 depicts the manageability of a hierarchical design.

Performance

Networks based on a hierarchical model are efficient in performance because of the use of advanced routing features such as *route summarization*. This feature enables

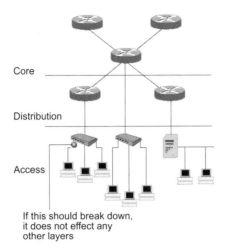

Core

Distribution

Access

If this should break down,
it does not effect any
other layers

FIGURE 3.4 Hierarchical design manageability showing different layers.

availability of small routing tables and faster convergence in large networks, thereby improving network service.

Cost

Because of its simple design, the hierarchical model offers benefits such as cost efficiency and ease of manageability. Bandwidth sharing with the layers of a WAN results in a decrease in costs and also requires less administration.

Scalability

The hierarchical model is easy to scale as compared to other network architecture models because it is divided into modules. As a result, modular expansion is easy and does not affect the rest of the internetwork. The hierarchical model can be compared to a production-based organization, ABCD Company, at which the manufacturing unit is connected to two wholesale outlets, which are connected to four retail outlets. Figure 3.5 depicts the hierarchical structure of the organization.

Now, ABCD Company is planning to expand its setup. The hierarchical structure has made it easy to expand the setup at any layer. Figure 3.6 depicts the structure of ABCD after it has undergone expansion.

Similarly, in the hierarchical model, you are not required to change the entire network topology. It is easy to scale and does not require any reconfiguration.

Speed

The hierarchical model makes it possible to provide high-speed switching of data between workstations and internetwork services. This is because all local traffic is

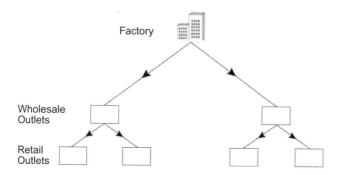

FIGURE 3.5   Hierarchical structure of ABCD Company showing different outlets.

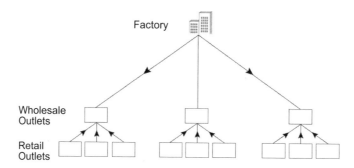

FIGURE 3.6   Hierarchical structure of ABCD after expansion.

resolved at the access layer, and only requests for enterprise services are forwarded to the core layer.

In addition to modularity, route summarization and fast-converging protocols increase the overall speed of network services.

FAULT-TOLERANT MODEL

A fault-tolerant network model is designed with failure handling capability as the primary objective. It operates uninterruptedly even when a network failure occurs. An organization concerned mainly with failure recovery uses this model for network design. Networks deployed for mission-critical applications such as space programs or telemedicine systems utilize this type of model. For example, a hospital, which stores critical and important data related to patients, opts for a fault-tolerant model. If a heart transplant surgery is in process and a network failure occurs, it could result in irreversible damage.

The functioning of a fault-tolerant system is divided into three steps:

Error detection: Identifying the section of the network where the failure has occurred

Switchover to the backup system: Ensuring that the network system performs normal operations and obtains acceptable results after one or more system components have failed

Creating log reports: Reporting errors to the operating system

To ensure that a network performs normal operations in the event of a network failure, redundancy is incorporated into the design. This design ensures that alternate systems are available as backups for all services in the fault-tolerant system.

When designing a network topology for a customer who has critical systems, services, or network paths, implement a fault-tolerant system combined with redundancy. For example, Mr. Davis, who is working as a network administrator in a large hospital, has identified a critical network connection between the blood laboratory on the ground floor and the operating room on the third floor. To protect the network against any single-point failures and to keep network segments functional during surgery hours, he placed two LAN switches in the topology to connect these two segments. When one switch fails, the other switch maintains network connectivity. The aim is to keep these network segments always functional, especially during critical surgical hours. Figure 3.7 depicts the two-switch concept between the laboratory and the operating room.

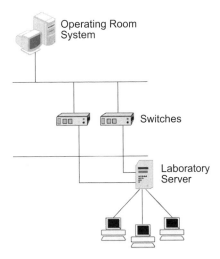

FIGURE 3.7 Two-switch concept between the laboratory and the operating room.

Types of Redundancy

You can incorporate different types of redundancy in a network design, one at a time or in different combinations to create a fully tolerant system. The types of redundancy are:

- Workstation-to-router
- Server
- Route
- Media

Workstation-to-router Redundancy

When a workstation sends data to a remote site, it needs to access a router. The workstation deploys different methods to locate the address of the router. The use of different methods to locate the router enables workstation-to-router redundancy. There are various ways to find the address of a router. The method used to locate a router depends on the protocols and network implementation. The different protocols and methods used to locate a router are as follows:

- Address Resolution Protocol (ARP)
- Explicit configuration
- Router Discovery Protocol (RDP)
- Hot Standby Router Protocol (HSRP)
- Internetwork Packet Exchange (IPX)
- Routing Information Protocol (RIP)
- Open Shortest Path First (OSPF)
- AppleTalk

Address Resolution Protocol (ARP)

Workstations use ARP to find remote sites. This protocol enables a workstation to establish an IP session by resolving the router address. Workstations send ARP frames to the router, which uses a *proxy ARP* to respond to an ARP request through the Data-link layer address of the router. The benefit of a proxy ARP is that the workstation does not have to be directly configured to the router address. Figure 3.8 illustrates the working of ARP.

Explicit Configuration

Explicit configuration is the most common method for a workstation to access a router. In this method, IP workstations are configured with the IP addresses of the default router. This is also called a default gateway. If the default router is unavailable, the workstation will establish the connection using a backup router. To do this, reconfigure the workstation with the address of the backup router. Figure 3.9 depicts an explicit configuration of a workstation to access a router.

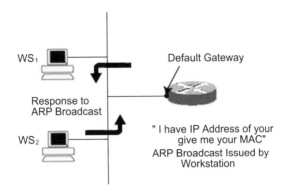

FIGURE 3.8 ARP functioning within different workstations using a router.

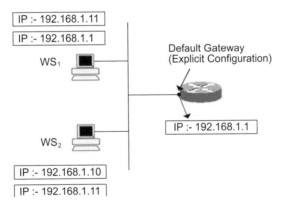

FIGURE 3.9 Explicit configuration of a workstation to access a router.

Router Discovery Protocol (RDP)

RDP allows connectivity between a router and workstation. Using this protocol, the workstation learns the address of the local routers by means of RDP. This protocol is an extension of Internet Control Message Protocol (ICMP). A router periodically multicasts an ICMP packet along with the RDP from all interfaces. This RDP contains the IP address of a router, which is broadcast to the workstation every 10 minutes. Figure 3.10 depicts the working of RDP and ICMP within a network.

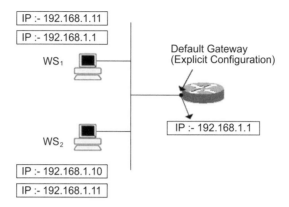

FIGURE 3.10   RDP and ICMP protocol connectivity within a network.

Hot Standby Router Protocol (HSRP)

HSRP provides redundancy for connectivity between the workstation and router by replacing the default router with an alternate one when the default router is unavailable. This replacement router is created as a virtual router using HSRP.

HSRP works by creating a virtual or phantom router. A phantom router does not exist physically but has a unique MAC and IP address. The workstation uses the IP address of the phantom router as the default gateway. This IP address is passed to the two physical routers used in HSRP. One of the two physical routers serves as primary (active) and the other as secondary (standby). When the active router fails to host the address of the phantom router, the standby router resumes data transmission.

The active router continuously sends "Hello" messages to the standby router. If a failure occurs and the active router stops sending "Hello" messages, the standby router takes over and becomes active. The standby router also uses the IP and MAC address of the phantom router, as was done by the active router. As a result, the workstation node sees no change.

The workstation is available with redundant routers in case of failure. Figure 3.11 depicts the functioning of HSRP.

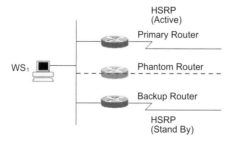

FIGURE 3.11   The mechanism of HSRP showing primary and secondary routers.

Internetwork Packet Exchange (IPX)

IPX workstations use the "find network number message" to detect a route to the server. A router responds to this message. If there is any break in connectivity, a similar message is sent out, and connectivity is re-established. Figure 3.12 depicts how IPX uses the "find network number message."

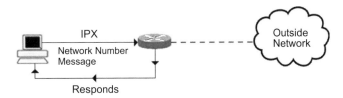

FIGURE 3.12   IPX using the "find network number message."

Routing Information Protocol (RIP)

RIP is a distance-vector protocol that uses hop count as the measuring unit and performs routing within a single Autonomous System (AS). A workstation runs RIP to learn about routers. The workstation sends RIP frames to locate the router; on receipt of the frame, the router sends a response to the workstation. RIP supports and governs access in conjunction with the Open Shortest Path First (OSPF) protocol (discussed in the next section). Figure 3.13 shows how RIP operates in a network along with OSPF.

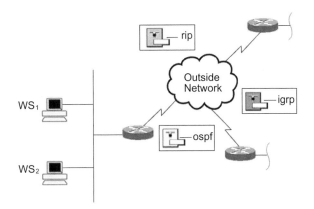

FIGURE 3.13   The operation of RIP in conjunction with OSPF.

Open Shortest Path First (OSPF)

OSPF is a *link-state* protocol, which calls for broadcasts from all the routers in the same hierarchical layer. The routers respond to OSPF with their own MAC address. OSPF then discovers the shortest open path using the information collected from the received broadcasts. The end result is that workstations can dynamically locate redundant paths to the router. Figure 3.14 depicts the working of the OSPF protocol within a network.

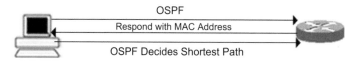

FIGURE 3.14   Working of OSPF using information from the received broadcasts.

AppleTalk

AppleTalk uses the Routing Table Maintenance Protocol (RTMP) to track the address of routers. RTMP sets subsequent connections. AppleTalk remembers the address of the router that sends the last RTMP packet.

Server Redundancy

Server redundancy refers to the creation of duplicate or backup instances of all types of servers. In some environments, all servers, including Dynamic Host Configuration Protocol (DHCP), name, Web, and broadcast, are fully redundant. In case of single-point failures, a backup server performs normal operation. This type of redundancy is important in situations when real-time data is transferred and processed. Server redundancy uses backup and hot swap failover support.

Redundant DHCP servers are implemented at the access layer. So the data traffic is divided between two DHCP servers. This redundancy of DHCP servers is important for network systems of organizations such as stock trading companies, which handle a large amount of data traffic.

File server redundancy is recommended in systems in which the organization's operational efficiency is data dependent. For example, in a financial organization in which account-related data and money transfers are critical, single point failures of file servers might result in loss of valuable information. To provide backup in case of network failure, this redundancy method is utilized to replicate data on two or more servers. Mirroring is a method used for file server redundancy. This involves synchronization of the two disks, which enables data replication between two servers. Figure 3.15 depicts server redundancy in a network. The network has a backup file server with mirrored disks.

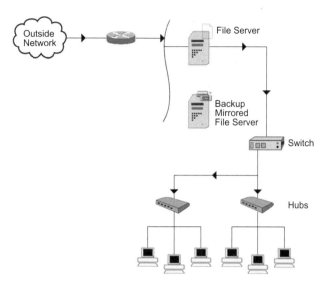

FIGURE 3.15 Server redundancy showing a backup file server with mirrored disks.

Route Redundancy

Route redundancy means that an alternate path is provided for a network in case of breakdown of the primary path. The network topology is designed such that multiple paths are available for the network. Therefore, route redundancy is a configuration pattern that provides backup link options to ensure that the network does not go down in case of link failure. The original and alternate routes should not be located next to each other because failure in the original route might impair the alternate route. Figure 3.16 depicts route redundancy in a network.

Designing route redundancy provides two features: load balancing and down-time reduction.

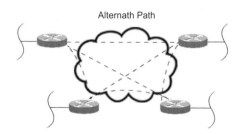

FIGURE 3.16 Route redundancy showing multiple alternate paths.

Load Balancing

One advantage of route redundancy is that it provides the ability to institute load balancing. *Load balancing* is the process of balancing the network load across parallel links such that a single link is not overloaded.

The main objective is that the bandwidth of all the parallel paths should remain consistent for supporting load balancing. If the path with a lower bandwidth is congested, the path with a higher bandwidth cannot be filled. This situation is called *pinhole congestion*.

To avoid pinhole congestion, you can also use IP routing protocols. Most of the IP routing protocols support load balancing by providing equal bandwidth. IP routing protocols can support a maximum of six parallel routes.

Downtime Reductions

Redundant routes also minimize network downtime. You can nullify the effect of link failures with the use of redundant meshed designs. A network may be designed as a full or partial mesh.

Full Mesh Design

A *full mesh* network design is used when a router is connected to all the other routers in the network, using n*(n-1)/2 number of links, where n denotes the number of routers. A full mesh network provides high redundancy and good performance but is expensive to implement. This is because a large number of links are required to implement a full mesh network, as shown in Figure 3.17.

This cost is prohibitive for networks with a large number of routers. Redundancy, when designed using meshed networks and with respect to hierarchical positioning, yields good results.

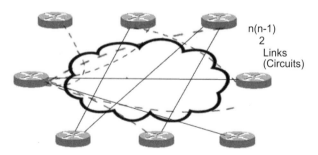

$$\frac{n(n-1)}{2}$$
Links
(Circuits)

FIGURE 3.17   A full mesh network providing high redundancy.

Partial Mesh Design

A *partial mesh* design is less redundant as compared to a full mesh design because only some nodes are in a full mesh scheme while other nodes are connected to only one or two nodes. Figure 3.18 depicts the partial mesh network design.

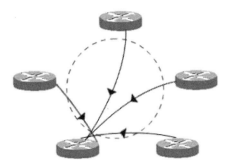

FIGURE 3.18   A partial mesh design with less redundancy in a network.

Dial-on Demand Routing Backup

Route redundancy is also implemented with the help of Dial-on Demand Routing (DDR) backup. Most of the organizations have multiple locations with their branches spread across a wide geographical area. To ensure continuous connectivity among the branches, redundant links are required. Maintaining dedicated redundant links is not cost effective, and as a result, ISDN is used as a backup for dedicated leased lines. ISDN becomes functional only when the leased line fails. This configuration is achieved only with the help of DDR. DDR activates ISDN when a primary line fails or when there is an increase in network traffic. Figure 3.19 depicts a DDR routing backup.

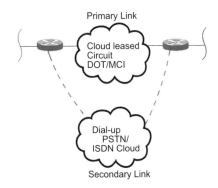

FIGURE 3.19   Dial-on demand routing backup showing primary and secondary links.

Media Redundancy

Redundancy of media is important in case of mission-critical applications. In a WAN, back-up links are often deployed for ensuring media redundancy. These back-up links are automatically activated when the primary link goes down.

In some networks, routers can have redundant links to each other. These redundant links provide alternate links for routers to communicate with each other. This reduces network downtime but may result in a large amount of broadcast among routers, causing a broadcast storm. Cisco protocols, such as the Spanning Tree Protocol (STP) and Spanning Tree algorithms reduce the looping effect of a broadcast and ensure that there is only one active path between two network stations. Figure 3.20 depicts primary and secondary routers on a network and their corresponding interconnections.

In Figure 3.20, the line between the primary and secondary server indicates that there is only one medium path for connectivity to the network systems. The unavailable lines for the media are shown by the dotted lines for the connections.

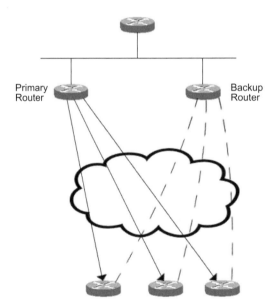

FIGURE 3.20 Primary and secondary routers showing their interconnections.

Types of Fault-tolerant Systems

The fault-tolerant system can be of two types, fully redundant or partially redundant.

Fully Redundant Fault-tolerant System

Within a *fully redundant* system, all the components are provided with backup, and the performance of this system is always predictable. A fully redundant system provides the recovery option even after multiple network failures. For example, consider an organization that provides redundancy to all network components—two hubs, two switches, and two servers. This ensures that the network performs normal operations without interruptions 24 hours a day and 365 days a year. Figure 3.21 illustrates a fully redundant fault-tolerant system.

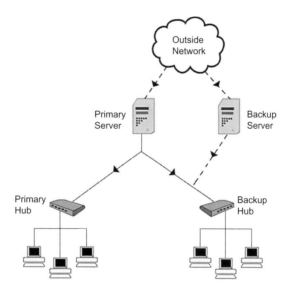

FIGURE 3.21   A fully redundant fault-tolerant system showing primary and backup servers.

Partially Redundant Fault-tolerant System

In a *partially redundant* system, only some modules are redundant. A partially redundant system is also called a fault-resilient system. This system is not provided with a complete backup of all the modules; and as a result, its performance may be affected. For example, the network of ABCD Inc. has many workstations that are attached to a hub. The hub is connected to two servers. This network is called a partially redundant system because it has a backup for the server, but there is no backup support for the hub. If one of the servers fails, the presence of the other server ensures normal operations. If the hub fails, normal functioning is interrupted. Most organizations opt for a partially redundant model because it is more cost efficient and easy to manage. Figure 3.22 depicts a partially redundant fault-tolerant system.

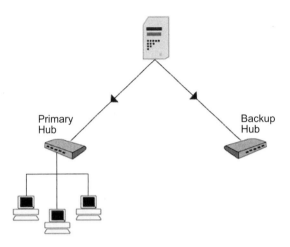

FIGURE 3.22 A partially redundant fault-tolerant system showing primary and backup hubs.

SECURITY MODEL

Security model architecture is designed with network security as the primary consideration. In this model, various network components are positioned to achieve the requisite level of security. This model is used by the organizations that offer extranet services without compromising the internal network security. For example, a defense organization needs to securely transfer sensitive data without the risk of the data being intercepted.

Organizations that offer network services without compromising their internal network use the security model. Do the following when designing secure topologies:

- Implement easy to manage security mechanisms in the network design.
- Isolate all the security services from the remaining system.
- Create awareness among users of security mechanisms implemented in the network design to avoid mishandling.
- Restrict access to sensitive areas of the network. For example, you can implement explicit permissions for access to the server.
- Implement special mechanisms to detect intrusions into the network.
- Avoid hampering of network performance through the secure model.
- Design the security model with the options of regular updates and modifications.

Mechanisms for Secure Topologies

For designing secure topologies, various mechanisms are implemented to enhance network security and performance, as follows:

- Basic built-in security devices
- Firewalls
- Hardening the network
- Intrusion detection system

Basic Built-in Security Features

The first step towards designing a secure network is to choose a network device with built-in security features that enable the network devices to detect and block intrusions, attacks, and malicious activities. Most network devices available today are manufactured with standard built-in security features. Examples of these devices are:

- Routers
- Switches
- Wireless-networking devices

Routers

The hardware market is currently flooded with broadband routers with built-in firewall and VPN capabilities. Routers also perform encryption and MAC address filtering. The firewall capabilities of these routers conceal your network, and the encryption techniques protect data from malicious attacks. While MAC address filtering grants or denies access permissions to the network, routers use Network Address Translation (NAT) to hide the network address when data is sent out.

Switches

To provide network security, switches are incorporated with enhanced security features such as built-in security wizards, which ensure that only authorized users access a specific segment of the network. In addition, switches also include ACLs that enable policy-based access to different network devices.

Wireless-networking Devices

Different wireless devices provide security based on *Wired Equivalent Privacy* (WEP) or *Wi-fi Protected Access* (WPA) encryption. The WEP and WPA encryption techniques are implemented with the help of an encryption key.

Firewalls

When a network is connected to an external network or the Internet, it is exposed to different security threats. Therefore, an organization has to create a logical boundary between its network and an unreliable network to allow only the right information to pass through. The use of a firewall makes it possible to create such a logical boundary.

A firewall is a hardware device or software that enables you to create a barrier between your network and any outside network. It protects the network from malicious activities and attacks, and filters traffic flowing through the network. Firewalls control access to network systems and protect them from unauthorized entities. They allow selective access to users and information on the network. Some firewalls block traffic while others permit only authenticated traffic. The data that flows through the network from an external source has to first pass through the firewall. An organization can set access controls on the firewall to limit the traffic entering the network.

Different functions of a firewall are:

- Providing centralized security
- Administering the traffic inflow
- Implementing access rights to sensitive systems
- Preventing intrusion of unreliable and potentially dangerous services into the network
- Maintaining detailed logs of users accessing the network
- Filtering traffic depending on end-user rights and privileges

Types of Firewalls

The two types of firewalls are hardware and software.

A *hardware firewall* is a small box; plug your computer into it, and the computer is protected against all types of security risks. The operating system is prehardened because there is a small computer within the hardware firewall box. Therefore, the prehardened operating system manages and filters traffic flowing through the firewall. The use of hardware firewalls also ensures port filtering and encryption. Thus, data flowing through the firewall is protected against virus attacks. A hardware firewall is deployed at a centralized position between the internal and external network.

A *software firewall* is a program that runs on a computer and protects it from security threats. Firewalls are host-based programs that can be implemented on different operating platforms, such as Windows 2000, Linux, or Unix. The software firewall monitors the incoming and outgoing data from the network. The software firewall automatically blocks or permits traffic into the network. In some cases, it may ask you to specify whether or not the traffic should be allowed to enter the network.

The functioning of software firewalls is based on certain sets of rules such as ACLs and security policies. These rules specify the type of traffic to be permitted.

For example, you can set a rule specifying that information coming from a particular Web site should be blocked.

Firewall Implementation

When implementing a firewall, you can keep certain considerations in mind:

- Determine the security policy on which the firewall implementation is based. The security policy specifies rules pertaining to incoming and outgoing traffic and denial of access.
- Identify the types of users for whom these security policies are to be implemented.
- When these points are implemented, the firewall is placed between insecure external networks and the secure internal network.

The *three-part firewall* system is a complex implementation of a firewall that uses the concept of a De-Militarized Zone (DMZ). The DMZ is a buffer between a secure corporate network and all external networks. This network is strengthened against outside network attacks with the creation of two DMZs, *DMZ I* and *DMZ II*. Splitting up the entire network into two such separate zones is useful when there are multiple servers present. Figure 3.23 depicts a network with various servers such as Web, DNS, and Telnet.

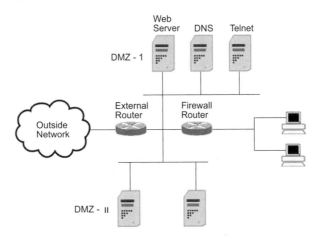

FIGURE 3.23   A network showing two de-militarized zones, DMZ I and DMZ II.

A DMZ contains the following components:

Isolated LAN: Represents the LAN segment between the external and internal firewalls

Internal Router: Serves as an internal packet filter between a corporate internetwork and an isolated LAN

External Router: Serves as an external packet filter between the isolated LAN and an outside network

Hardening the Network

To protect the network against attacks, you can implement a process called *hardening the network*. This refers to the periodic updating of protection mechanisms such as patches and vendor updates. You can strengthen the network using these methods:

Hardening of ports: Open ports are vulnerable to security attacks. Therefore, keep a check on all the ports that are activated on your network interfaces. There are two methods to check for activated ports on your network. The first method is to check the network stack and the other is to use port scanner utilities.

Enabling and disabling services: To reduce security attacks, disable services that are currently not in use. For example, if your network uses the attached scanner only periodically, set the scanning service of the Network Operating System (NOS) not to run when the scanner is not in use. Each operating system provides a service configuration tool or a set of commands to enable and disable services. Although any network service can lead to an insecure system, there are services such as Telnet, FTP, HTTP, and SMTP that have an inherent lack of security. You need to avoid enabling such services.

With the help of hotfixes: To fix bugs, apply hotfixes. Vendors regularly issue hotfixes, but these are not always stringently tested. You should install hotfixes on those workstations that face the threat of an attack and not on every workstation of the network. Also, upgrade the workstations whenever a service pack is available. Service packs are thoroughly tested by the vendor and minimize the risk of an attack.

ACLs: ACLs are filters specified in a firewall at the router level. ACLs specify the criteria for permitting or restricting data to pass through the network. You can specify one or more criteria for a particular ACL.

You also need to frequently update the NOS to ensure that no vulnerability exists in the system. To ensure that the operating systems can be updated without reinstalling, there should be a built-in support for upgrading them. When new problems arise, vendors issue corresponding software patches. These patches correct the vulnerabilities by making direct changes to the NOS and system configuration.

Although software patches reduce the vulnerabilities of the operating system, they may reduce the performance of the network. In addition to the patches, software updates are also used to harden the NOS. Updates (also called firmware) are programs installed into *Programmable Read-Only Memory* (PROM). These updates fix bugs and improve network performance. Firmware is a permanent part of a computing device. Vendors of devices distribute firmware whenever there is requirement for upgrading PROM.

An organization needs to secure the file system for hardening the NOS. For this, create encrypted disk volumes. The data in file systems is divided into small disk volumes, which are encrypted to secure the data. Another way of securing the file system is to assign access rights to authorized users only.

Intrusion Detection System

An intrusion is an attempt to break into a system to steal confidential data. VPNs and firewalls offer perimeter (boundary) security, but these have several limitations. Boundary security systems do not monitor authorized users when they are a part of the network. This may pose a security threat to the organization. For example, an authenticated dial-up user or an authorized person in an organization can misuse services or attempt to probe the network for the purpose of pursuing malicious activities.

In addition, when the network is connected to an external network through unsecured modems, a firewall cannot detect malicious traffic moving within the network, leaving the network open for attack.

You need a system to detect and counter probable intrusions from outside and within the network because security systems deployed at the boundary of the network may not provide sufficient protection. Therefore, organizations deploy *Intrusion Detection Systems* (IDSs) to identify and stop potential attacks and misuse of network from users—outside and within the network.

IDS is an application or process that continuously monitors a network to detect unscrupulous activities that indicate network abuse, misuse, or malicious attacks. The IDS monitors network activities by sniffing packets that flow in a network. It constantly works in the background and notifies the administrator when it detects any suspicious activity.

In addition, IDS monitors all actions on the network. When a new malicious action is detected, it is logged and marked as suspicious. The IP address of this action is tracked and monitored and finally sent to the administrator. Figure 3.24 depicts an IDS installed at the first point of entry on a network.

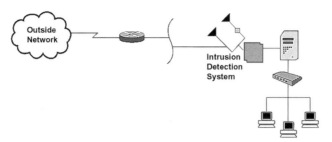

FIGURE 3.24 Network with an installed
IDS for internal and external protection.

SUMMARY

In this chapter, we learned about the Cisco Hierarchical model, which was designed on the premise that the volume and patterns of network traffic significantly impact network performance. We also reminded you about the three layers of the Cisco hierarchical model and their functions and benefits. In the next chapter, we move on to provisioning hardware and media for LAN.

POINTS TO REMEMBER

■ The three layers of the Cisco hierarchical model are core, distribution, and access.

■ The core layer enables high-speed switching of large volumes of traffic between end-users and enterprise services.

■ The distribution layer controls traffic and movement of data between the core and access layers and implements network policies.

■ The access layer connects all end users to the LAN, ensures that local server traffic is not forwarded to the wider network, and controls user and workgroup access to internetwork resources.

■ The fault-tolerant network model is designed to operate uninterruptedly even in the event of a network failure.

■ Workstation-to-router redundancy involves the use of different protocols, such as ARP, explicit configuration, RDP, HSRP, IPX, RIP, OSPF, and AppleTalk to locate the router when the workstation needs to send data to a remote site.

■ Server redundancy refers to the creation of duplicate or backup instances of all types of servers.

■ Route redundancy signifies that an alternate path is provided for a network in case of a breakdown of the primary path.

- Load balancing is the process of balancing the network load across parallel links, such that a single link is not overloaded.
- Full mesh is used when a router is connected to all the other routers in the network, using n*(n-1)/2 number of links, where n denotes the number of routers.
- Partial mesh is less redundant as compared to a full mesh design because only some nodes are in a full mesh scheme while other nodes are connected to only one or two nodes.
- DDR enables route redundancy by activating ISDN when a primary line fails or when there is an increase in the network traffic.
- Media redundancy is very important for mission-critical applications; back-up links are automatically activated when the primary link goes down.
- The fully redundant fault-tolerant system ensures that all the components on a network are provided with backup, and the performance of the entire system is always predictable.
- The partially redundant fault-tolerant system is not provided with a complete backup of all the modules; and as a result, its performance may be affected.
- The security model is designed with network security as the primary consideration, and the various network components, such as firewalls and IDS, are positioned to achieve the requisite level of security.

4 Media and Hardware Selection for Provisioning LAN

IN THIS CHAPTER

- Provisioning Hardware for LANs
- Provisioning Media for LANs
- Cisco LAN Equipment

The first step towards designing a LAN is selecting the appropriate hardware and media. An incorrect assessment of the hardware and media requirements for the proposed network can result in poor network performance and significant financial losses. To select the correct hardware and media, you need to have in-depth knowledge of the types, features, functions, and specifications of the hardware and media used for designing a LAN.

PROVISIONING HARDWARE FOR LANs

A LAN is a group of computers connected to each other with the help of cables and wires within a limited geographical area. These devices are designed to enable data transmission across small distances, making these suitable and cost efficient for implementing LANs. The types of hardware devices required for provisioning a good LAN are:

- Repeaters
- Hubs
- Bridges
- Routers
- Switches

Repeaters

Repeaters are Layer 1 devices that connect different media segments of an extended network. Repeaters receive signals from one segment and amplify and retransmit them to the other segment of the network. Repeaters are used to connect wire segments in a network so that limitations of segments due to distance do not hinder the process of data transmission. However, a repeater is an unintelligent device because it only repeats the signal without understanding the source or destination of the data.

Repeaters are used mainly on Thinnet and Thicknet (coaxial cable) networks.

Repeaters can be separate devices or incorporated into a concentrator (discussed later in this chapter). You can use them to regenerate analog or digital signals that are distorted by transmission loss. While analog repeaters only amplify a signal, digital repeaters reconstruct a signal near to its original quality.

Repeaters are used when you need to extend your network beyond 100 meters (328 feet). However, you can use only five repeaters in a series in order to extend the cable length for a single network. For example, consider a LAN using a star topology with Unshielded Twisted-Pair (UTP) cabling. You can use a multiport active repeater to connect each node through a twisted-pair cable. The length limit for a UTP cable is 100 meters. The repeater amplifies all the signals that pass through it, allowing the total length of the network cable to exceed the 100-meter limit. As a result, you can prevent signal deterioration with the use of long cables and a large number of connective devices.

There are some limitations to using repeaters:

- They are incapable of understanding frame formats, that is, they are unable to distinguish between a valid frame and other electrical signals.
- Unlike bridges/switches, they do not have physical addresses.
- They do not perform filtering and traffic processing.
- Only four repeaters can be included between any two Ethernet stations.

In a data network, a repeater can relay messages between subnetworks using different protocols or cable types. According to the IEEE 802.3 specifications, a network using repeaters between communicating stations (PCs) is subject to the 5-4-3 rule of repeater placement on the network. The 5-4-3 rule states that a maximum of five segments should be connected through four repeaters/concentrators between two communicating stations. If they use coaxial cable, only three of the five segments should be populated (trunk) segments. In addition, no fiber segment should exceed 500 meters (1640 feet).

This rule does not apply to other network protocols or Ethernet networks in which fiber optic cabling or a combination of a fiber backbone with UTP cabling is used. If there is a combination of fiber optic backbone and UTP cabling, the rule is translated to a 7-6-5 rule. According to the 7-6-5 rule, a maximum of 7 segments should be connected through 6 repeaters/concentrators between two communicating stations. Only 5 segments should be populated in a case of the 7-6-5 rule.

NOTE

The 5-4-3 rule was created when the Ethernet, 10Base5, and 10Base2 were the only types of networks available. This rule applies to the shared-access Ethernet backbones. A switched Ethernet network should be exempt from the 5-4-3 rule because each switch has a buffer to temporarily store data, and all nodes can access a switched Ethernet LAN simultaneously.

Hubs

Like repeaters, hubs are Layer 1 network devices. The function of hubs is to receive signals from a segment, amplify (except in the case of passive hubs), and broadcast them to all devices on the network including the broadcasting device. Hubs regenerate and retime network signals. This is performed at the bit-level to a large number of hosts (for example, 4, 8, 24) using a process called *concentration*. When a packet arrives at a port, the packet is copied to other ports. As a result, all segments of the LAN can see the packet.

A hub is known as a multiport repeater, that is, it contains multiple ports, in which data arrives from one or more directions and is forwarded in other directions in a network. Figure 4.1 depicts the placement of a hub in a network.

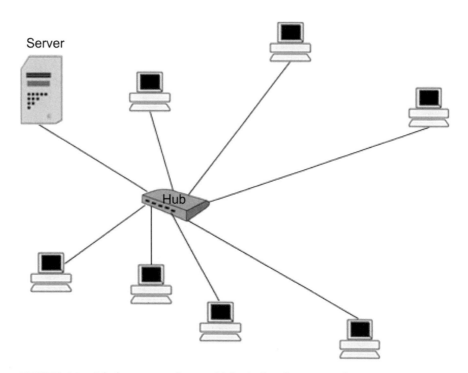

FIGURE 4.1 A hub connected to multiple devices in a network.

There are three types of hubs: active, passive, and intelligent. Each of these hubs has a specific set of features and offers different levels of services.

Passive Hubs

A passive hub serves as a medium for data to move from one device or segment to another. This hub receives all incoming packets on a single port and rebroadcasts them across all ports. It does not enhance the performance of a LAN or enable it to identify performance bottlenecks. Passive hubs usually have one 10Base2 port connected to each LAN device through RJ-45 connectors.

Active Hubs

An active hub amplifies a weak signal of an incoming packet before broadcasting it to other segments in a network. In addition, it also repairs damaged data packets and retimes the distribution of other packets. Therefore, an active hub boosts the signal as it travels from one node to another. Active hubs are more expensive than passive hubs and are available in different configurations with various ports.

Intelligent Hubs

An intelligent hub is also known as a stackable hub because it is possible to place multiple units on top of each other. Such configuration occupies less space. Intelligent hubs are able to transmit data at the rate of 10, 16, and 100 Mbps to desktop systems using standard topologies, such as Ethernet, Token Ring, or FDDI. They are also called manageable hubs because intelligent hubs provide remote management capabilities via Simple Network Management Protocol (SNMP) and Virtual LAN (VLAN) support.

A passive hub is called a concentrator while an active hub is known as a multiport repeater.

NOTE

Bridges

When networks grow in size, they are often divided into small LAN segments to efficiently manage the network traffic. A bridge is a Layer 2 device designed to connect two LAN segments. The purpose of a bridge is to filter traffic on a LAN, keep the local traffic restricted, and yet allow connectivity to traffic that has been directed to other segments of a LAN. A bridge monitors packets as they move between segments, keeping track of the MAC addresses that are associated with various ports. It is capable of managing traffic flow better as it starts to gather information about the nodes connected to each network.

A bridge identifies the local and the global traffic using the MAC address. Each networking device has a unique MAC address on the NIC. The bridge tracks MAC addresses on each side and makes its decisions based on this MAC address list. Figure 4.2 depicts the utilization of a bridge in two departments of an organization and within those departments.

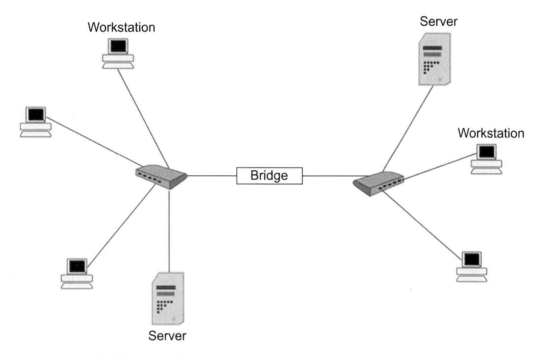

FIGURE 4.2   A bridge connecting two LAN segments.

Routers

Routers are Layer 3 devices. Routers are multipurpose devices that segment a network, limit broadcast traffic, and provide security, control, and redundancy between individual broadcast domains.

In an internetwork, a router routes the data packet from the source node to the network address of the destination node. For routing data packets, a router creates and maintains a table of various available routes. This helps evaluate the best possible route for the packet and improves network performance.

Routers perform a variety of functions, such as:

■   Creating and maintaining routing tables for each Network layer protocol. The tables are created either statically using manual configuration or dynamically using distance-vector or link-state routing protocols.

- Performing packet forwarding by identifying a specific protocol in the packet and forwarding it to the extracted destination address.
- Controlling and segmenting the network into individual collision domains within a broadcast domain.
- Performing specialized services such as policy-based routing, traffic flow management over the network, hierarchical addressing, and load balancing.

Cisco is one of the industry leaders in the manufacture of routers and offers several series of routers:

- Cisco SOHO 70 Series
- Cisco 800 Series
- Cisco 1700 Series
- Cisco 2600 Series
- Cisco 3600 Series
- Cisco 3700 Series
- Cisco 7100 Series
- Cisco 7200 Series
- Cisco 7300 Series
- Cisco 7400 Series
- Cisco 7500 Series
- Cisco 7600 Series
- Cisco 10000 Series
- Cisco 10720 Series
- Cisco 12000 Series
- Cisco MC3810 Multiservice Concentrator
- Cisco AS5200/AS 5300/SA5800 Access Servers

The Cisco 800, 1600, and 2500 series are access routers, which allow access to individual users or small user groups. Models 7000, 7500, and 12000, also known as Gigabit Switch Routers (GSR), are powerful and modular routers used as backbone routers for high-end routing.

Switches

Switches are Layer 2 devices. Switches are special-purpose devices designed to address issues pertaining to network performance and bandwidth availability. A LAN switch is a device that channels data packets received from the source to the destination port. It scans the packet or the frame, and using the MAC addresses, it switches the packet to its desired destination. In addition, it provides higher port density than bridges. When a hub receives a digital signal from a node, the hub sends the signal to all ports attached to it while a switch sends the signal only to the specific port where the destination MAC address is located. Figure 4.3 depicts a sample configuration for LAN switches.

This figure depicts how Node A of Port 1 sends information to Node B of Port 2, directly through LAN switching, without sending it through other ports. Nodes A and B, and Nodes C and D can also communicate through switches without disturbing the other port.

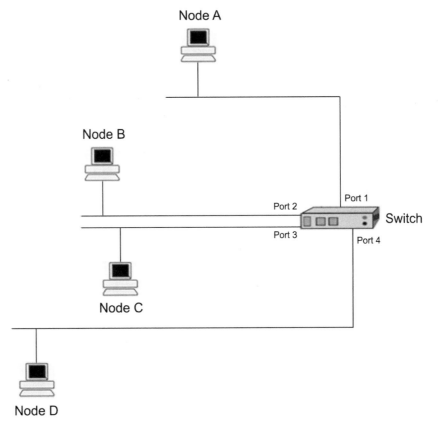

FIGURE 4.3   The configuration of a LAN switch.

Cisco is also an industry leader in the manufacture of switches, offering the following series of switches:

Cisco Catalyst 6500 series: Used in campus networks for multiplayer switching solutions

Cisco Catalyst 3750 series: Used in midsized organizations and enterprise branch offices

Cisco Catalyst 3550 series: Used in enterprise networks to provide high security and enhanced performance

Table 4.1 lists major differences between routers and switches.

TABLE 4.1   Differences between Routers and Switches

Routers	Switches
Focus on limiting broadcast traffic and providing security, control, and redundancy between separate broadcast domains	Focus on bandwidth availability for the network users resulting in enhanced network performance
Operate on Layer 3 of the OSI reference model	Operate on Layer 2 of the OSI reference model
Perform policy-based routing, and as a result, avoid occurrence of errors, such as packet looping and multiple replications	Do not perform policy-based routing, and as a result, packets are susceptible to looping and packet replication
Use routing protocols such as Open Shortest Path First (OSPF) and Enhanced Interior Gateway Routing Protocols (EIGRP), resulting in fast network convergence	Take topology decisions locally, leading to longer network convergence
Support firewalling and perform broadcasts and multicasts	Do not perform broadcasts and multicasts

PROVISIONING MEDIA FOR LANs

After the devices required for constructing or designing a LAN are streamlined, decide the appropriate media and standards for data transmission. This section discusses provisioning media for a LAN.

Ethernet

Ethernet is a network technology used for transmitting data packets to different nodes in a network. Xerox first developed Ethernet in the 1970s. The first version of Ethernet was released in 1980 as a joint venture of Digital Equipment Corporation, Intel, and Xerox (DIX). This standard, for the first time, set10Base5 or thick

Ethernet with Carrier Sense Multiple Access/Collision Detection (CSMA/CD) as the protocol for data transmission.

In 1983, the Institute for Electrical and Electronic Engineers (IEEE) formed a committee called 802 to create rules, protocols, and hardware standards for Ethernet cards and cables. The benchmark associated with the Ethernet is 802.3. The current IEEE 802.3 has standards for thick/thin, twisted pair, and optical fiber with speeds of 10 Mbps, 100 Mbps, and 1000 Mbps (used in LAN Ethernets). The 802.3 standards also set the standards for hardware (network cards and cables) for Ethernet networks. All current implementations of Ethernet follow the IEEE 802.3 standards.

The advantages of using the Ethernet are:

- Low network implementation costs
- Easy to implement compared to other options, such as Token Ring and FDDI
- Easy network management and maintenance

Some of the limitations of using Ethernet are:

- Cabling quality limits the number of nodes attached and the distance between each node in an Ethernet network segment. The cables are susceptible to electrical interference. As a result, to ensure optimum clarity for data transmission, the distance covered between machines has to be small.
- The number of nodes connected in an Ethernet segment is also limited because of the use of the CSMA/CD protocol. This protocol allows only a single mode of transmission among nodes. As a result, the fewer nodes on a network the higher the transmission speeds.
- Ethernet LANs also face the problem of network congestion in the case of bigger Ethernet LANs. An increase in the number of nodes results in a higher collision rate and delays in data transmission.

Ethernet Topologies

A topology determines how all nodes in a network are aligned or positioned with respect to each other. A topology can be either logical or physical. There are different topologies in a network such as bus, ring, star, and tree. In an Ethernet network, all nodes are aligned using either the bus or star topology. This section discusses Ethernet topologies in detail.

Bus Topology

In the bus topology, all nodes are connected to form a single line or bus structure. This structure involves data transfer from a node to all other nodes in the network

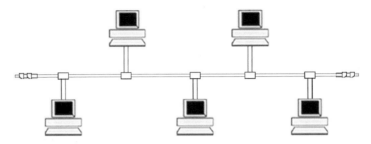

FIGURE 4.4 An Ethernet bus topology.

connected through a single cabling system. The earliest Ethernet networks used a linear bus structure. Figure 4.4 depicts an Ethernet bus topology.

This type of topology is not in use currently because of insufficiency in handling cable problems.

Star Topology

In a star topology, all the systems/nodes are connected to central devices, such as hubs and repeaters. All data packets are transmitted through this central device. Figure 4.5 depicts a star topology.

Most modern Ethernet networks are implemented using star topology, the backbone being either coaxial cable or fiber optic with a point-to-point link.

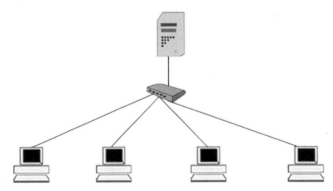

FIGURE 4.5 Multiple nodes connected to a central device forming a star topology.

Ethernet Operations

In an Ethernet network, each node operates as an independent unit. Such a unit is connected to shared media and interconnected with physical hardware and a set of protocols. Every node in an Ethernet network transmits data across the Ethernet through data packets called frames. The sublayer that transmits information is Media Access Control (MAC) sublayer. MAC resides in each node and determines the connectivity of each unit with the shared media. It is based on the CSMA/CD protocol for data transmission. All computers connected together in an Ethernet network communicate with each other through the MAC sublayer.

CSMA/CD

CSMA/CD transmits data from the source to the destination node in the form of frames. Each node sends the frame only after checking for availability of the network path. If another node also transmits a frame at the same time, it results in a collision, which causes the sent frames to be discarded. When this happens, each node continues to send the information after random intervals until the transmission is complete.

In CSMA/CD, "CS" stands for "Carrier Sense," which means that before sending data frames to different nodes or stations, the sender will check if other nodes are also sending data at the same time. Only if the Carrier Sense finds that the LAN is not transmitting any data or is in the idle state does the node transmit the frame. "MA" stands for "Multiple Access," which means that every node in the Ethernet Network is connected through the same cabling system (coaxial cable or optical fiber) to form a single path and is accessible simultaneously. "CD" stands for "Collision Detection," which means that the nodes also detect if any collisions had occurred in transmission.

Ethernet MAC Sublayer

The Media Access Control (MAC) sublayer is the one that transmits information through an Ethernet network. It is compatible with other Layer 2 protocols such as Token Ring. The MAC sublayer is specified at the Data-link layer. In Ethernet-based networks, the functions of MAC are:

- Formatting and creating data into packets. This process is called *data encapsulation*.
- Receiving frames from network devices and intelligently discerning the frame end and beginning.
- Managing and manipulating MAC addresses.

Ethernet Frame Structure

The Ethernet frame format structure consists of elements contained in the frame format or a part of the frame and the extensions and techniques added to guarantee safe transmission and receipt of transferred data.

Frame Format

For all MAC implementations, IEEE 802.3 has set a basic data frame format. A frame is a sequence of bits that contains addresses of the destination and source nodes. The elements of the frame enable successful data transmission across the network. A frame also ensures that the data has reached the destination node unhampered. The main elements of a frame format are as follows:

Preamble (PRE): Contains 7 bytes and is a combination of alternating 1s and 0s This sequence helps detect the presence of the frame (data) on the source and destination nodes after the data reaches its destination.

Start-of-Delimiter (SoD): Contains a sequence of 8 bits and a bit configuration of 10101011 that indicates the start of the frame.

Destination Address (DA): Contains 6 bytes, and as the name suggests, it identifies which node should receive the frame. The left-most bit in the DA also helps identify whether the address mentioned is an individual address (denoted by 0) for a single node or a group address (represented by 1) for a group of nodes. The second bit indicates whether DA is administered globally (0) or locally (1). The remaining 46 bits have a unique value that identifies a single node, a specified group of nodes, or all nodes on the network.

Source Addresses (SA): Contains 6 bytes and identifies the node that sends the packet. This address is always the same for a single transmitting device with multiple destinations, and the left bit is always 0.

Length/Type: Contains 4 bytes and indicates either the length or number of MAC-client data bytes that are contained in the data field of the frame. It also determines the frame type ID.

MAC client data: Contains the data to be transferred from the source to the destination node. The size of the frame should vary between 64 and 1500 bytes. If the size is less than the desired size, padding is used to bring the size to the minimum length. Padding refers to the appended data bytes required by the field to increase the length to the minimum required level.

Frame Check Sequence (FCS): Consists of 4 bytes and contains a 32-bit Cyclic Redundancy Check (CRC) value. When the source node assembles the data to be sent to the destination node, it performs a CRC on all the frame bits. When the check is complete, the node sends the data. In a similar way, a CRC is performed to check the frame when it reaches the destination nodes. After CRC, if the value is not the same, the frame is sent back along with an error report. The FCS is generated over the DA, SA, Length/Type, and Data fields.

Interframe Gap

Ethernet devices are required to allow a minimum interval between frame transmissions in order to prevent collisions. This interval is called the InterFrame Gap

(IFG). It is a kind of preparatory phase for the recipient node before it receives another frame. Table 4.2 lists the IFG for various types of Ethernets.

TABLE 4.2   IFGs in Ethernet Devices

Type of Ethernet	IFG
10 Mbps	9.6 microseconds
100 Mbps	960 nanoseconds
1 Gbps	96 nanoseconds

Frame Extension

A *frame extension* is an extension field added at the end of a frame set to increase the length of the frame to 512 bytes. Frame extensions are used in Gigabit Ethernet networks in the half-duplex mode. Its functioning is similar to Packet Assembler/Disassembler (PAD).

NOTE

Gigabit Ethernet is an Ethernet standard also known as 1000BaseT. Its theoretical speed of transmission is 1 Gbps.

Frame Transmission

We just looked at the frame format. We will now look at how a frame is transmitted and the mode to transmit the frame. Various terms associated with frame transmission are:

Slot time: Refers to the total or the minimum time taken by a node to wait before retransmitting the data. This is specified as 512 *bit times* (the time taken for transmission of one bit) for Ethernet networks operating at 10 and 100 Mbps and 4096 bit times for a Gigabit Ethernet network. This is useful for the transmitting medium/node to detect collisions. The minimum transmission time for a complete frame should be at least one slot time, and the time required for collision propagation for all stations on the Ethernet should be less than one slot time.

Propagation delay: Refers to the time taken by each signal for notifying the delay or traveling between stations present in the network.

The data frame in an Ethernet network is transmitted as a sequence of pulses or data symbols. During data transmission, it is reduced in size or is changed. When the data reaches the recipient, the recipient extracts the data symbols and corrects the frame before retransmitting the data.

Early Ethernet implementations used the Manchester encoding method. With advancement in Ethernet standards, this method was dropped because it was not compatible with Ethernet standards that used higher data transfer rates. The latest Ethernet standards use one of the following new signal-encoding schemes for transmitting data:

Data scrambling: Bits are rearranged without affecting the data frame. For example, 0s are changed to 1s. This rearranging or scrambling helps increase the transition density and provides easier clock recovery.

Expanding the code frame: The codes for data and control symbols are added or modified. For example, a field padding is used to increase the size of the frame to its minimum size requirement of 46 bytes. This expansion of the code frame helps detect errors during the transmission process and at the destination.

Forward error-correcting codes: Redundant information is added to the sent data stream. As a result, errors can be corrected during frame reception at the destination.

In an Ethernet network, data communication between systems or applications can be in one of two modes, half duplex or full duplex. The original Ethernet mode of operation used the half-duplex mode, which uses the CSMA/CD system and repeater hubs. The full-duplex Ethernet is based on switching hubs and does not use CSMA/CD. The two modes are discussed in detail next.

Half Duplex

The half-duplex mode of data transmission means that data cannot be transmitted and received at the same time. If a node transmits data at a given period, it can receive data only when the transmission is complete. The half-duplex mode uses the CSMA/CD algorithm for transmission. The sequence of steps for a half-duplex mode is:

1. Before transmitting a frame, a node monitors the network for the presence of a transmission carrier.
2. If a frame has already been transmitted, the node defers its own transmission and waits until the network enters the idle state before sending its data.
3. If the node senses the idle state of the network, it immediately starts sending the frame. In addition, the node monitors the network for collisions.
4. If a collision occurs, the node immediately sends a jam sequence message to notify all nodes of the frame transmission delivery error.
5. The node waits for a random period using a random number generator before sending the data frame. This process of waiting is called *back off*.

6. If the node encounters a collision again, the random delay period is increased until the data is successfully transmitted to the recipient node.

Full Duplex

The full-duplex mode of data transmission allows simultaneous data transfer between two nodes. This means that a node can send and receive data at the same time, resulting in enhanced network performance.

The primary drawback of the half-duplex mode is that it is capable only of either sending or receiving data at a given point in time. This means that stations can either send or receive data but not perform both simultaneously. In order to overcome this shortcoming, the IEEE 802.3x standard introduced the concept of full-duplex mode of transmission in Ethernet.

An important feature of full-duplex mode is *link aggregation* or *trunking*, which provides better or dedicated bandwidth and link availability among nodes. These benefits are achieved by combining all the physical links to form a single logical link. Link aggregation adds a new layer between the MAC sublayer and higher layer protocols. It aggregates all the ports to form a single logical link with one MAC address. Therefore, the frame works on the addresses present in the frame when it passes through this added layer.

Link aggregation enables load balancing by distributing network traffic across multiple links. This process also provides redundant links such that a single point failure of a link does not impact transmission.

The advantages of a full-duplex mode of communication are:

- Increased throughput because all nodes can simultaneously send and receive data.
- Increased link efficiency because the possibility of collisions is removed.
- Ensuring propagation of collisions to all stations within the required 512 bit times because there is no limitation in segment lengths as in the half-duplex Ethernet. This helps in collision detection. For example, 100BaseFX is limited to a 412-meter segment length in the half-duplex mode but may support segment lengths as long as 2 km in the full-duplex mode.

Ethernet Physical Layer

All Ethernet technologies and devices are implemented at the Physical and Data-link layers of the OSI model. The Data-link layer is logically subdivided into the MAC and LLC sublayers, and the Physical layer is directly linked to the physical cabling, such as 10BaseT used in Ethernet networks. Figure 4.6 shows the Ethernet model mapped to the OSI reference model.

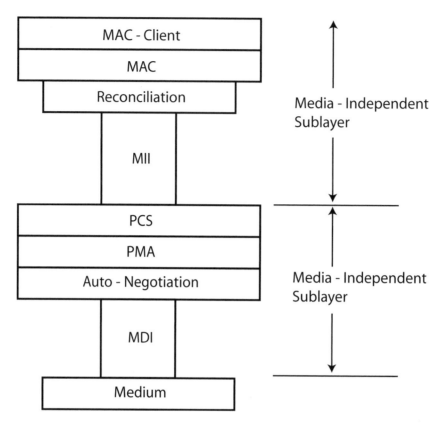

FIGURE 4.6   The Ethernet model mapped to the OSI model.

The Ethernet model consists of the following media-independent and media-dependent sublayers:

Reconciliation sublayer: Serves as a logical interface between the MAC and the media-dependent sublayers.

Media-Independent Interface (MII): Serves as a connecting link between the MAC sublayer and Physical layer devices. An example of MII is cable.

Physical Coding Sublayer (PCS): Performs encoding, multiplexing, and synchronization of the outgoing data frame or data symbols. This sublayer also performs decoding and demultiplexing of the data frame at the destination.

Physical Medium Attachment (PMA): Contains data frame transmitters and receivers and the recovery logic for the data stream.

Medium-Dependent Interface (MDI): Serves as a cable connector between two ends (transceivers and links).

Auto-negotiation layer: Operates on the NIC and exchanges information and negotiations about the mode of communication between two links/nodes.

Ethernet Cabling Medium

The earliest implementations of Ethernet used 10Base5 and 10Base2. Today, most organizations are upgrading to 10BaseT and 10BaseF cabling media.

10Base5

10Base5 is the Ethernet system that carries data at the rate of 10 Mbps using thick coaxial media. This system is used mainly in a bus architecture. The maximum segment limit for thick coaxial cable is 500 meters (1640 feet), and the maximum number of segments per attachment is 100. The main elements of 10Base5 are:

Terminator: Installed at each end of the cable.

Multistation Access Unit (MAU) or transceiver: Placed at an interval of 2.5 meters (8 feet 2-1/2 inches) and reduces the signal reflection that degrades the quality of the signal. The maximum distance limit between an MAU and a segment is 100 meters (328 feet).

Attachment Unit Interface (AUI): Placed between the NIC and the Ethernet cable and provides power to the MAU. It also carries signals between the Ethernet interface and the MAUs.

Ethernet interface: Board installed or built into Digital Terminal Equipment (DTE). The Ethernet cable is attached to this interface.

Table 4.3 lists the specifications for 10Base5.

TABLE 4.3 10Base5 Specifications

Characteristic	Specification
Transfer rate	10 Mbps (full-duplex not supported)
Topology	Bus
Maximum segment length	500 meters (1640 feet)
Permits DTE connection	Yes
AUI length	50 meters (164 feet)
Composition	Thick, heavy, and inflexible
Number of segments	100
Types of connectors	N-type coaxial, barrel connectors, and terminators
Scalability	Easy to scale
Frame transmission rate	0.65c* (195,000 km/sec or 121,000 miles /sec) where c is the speed of light

10Base2

10Base2 is an Ethernet system that transmits data at the rate of 10 Mbps using thin coaxial media. This is also known as the thin Ethernet or Cheapernet. This system is based on the bus topology where the maximum segment limit is 200 meters (656 feet), and the maximum transceivers per segment is 30. In addition, there should be a spacing of 0.5 meters (about 1 foot 7 inches) between two nodes. The transceivers are connected through a BNC tree and not through tapping as in the case of coaxial Ethernet. This BNC tree involves connectors directly connected to Ethernet cards and supports a daisy chain topology, in which the cable segment of a node is directly routed to another. The end of a 10Base2 coaxial segment is terminated with a BNC 50-ohm terminator. As a safety measure, a wire connects the segment to a ground to avoid grounding problems. Table 4.4 lists the specifications for 10Base2.

TABLE 4.4   Specifications for 10Base2

Characteristic	Specification
Transmission rate	10 Mbps
Composition	Thin, flexible cable
Topology	Bus
Segment length	185 meters (606.9 feet)
Maximum transceivers/segment	30
MAU	Present in the interface card
Connectors	BNC T and barrel connectors
Signal encoding	Manchester encoding
Frame transmission rate	0.59c (177,000 km/sec or 110,000 miles/sec where c is the speed of light)

10BaseT

10BaseT is the most common version of Ethernet system. It transmits data at the rate of 10 Mbps using voice-grade twisted pair media.

10BaseT is based on the star topology, and all the cable connections are point-to-point. Point-to-point connections result in maximum use of transceivers, which are attached to each end of the cable. Table 4.5 lists specifications for 10BaseT.

TABLE 4.5 10BaseT Specifications

Characteristic	Specification
Transmission rate	10 Mbps
Cable composition	UTP cabling
Segment length	100 meters (328 feet)
Maximum transceivers/segment	2
Connectors	RJ-45 style modular jack (8-pins)
Topology	Star
Frame transmission rate	0.66c (198,000 km/sec or 123,000 miles/sec where c is the speed of light)

10BaseF

10BaseF is an enhanced version of the Fiber Optic Inter-Repeater Link (FOIRL) standard. It uses fiber optic media for data transmission with a speed of 10 Mbps. The 10BaseF Ethernet system includes the three following types of fiber optic segments:

10BaseFiberLink (10BaseFL): Transmits data over two fiber optic cables at 10 Mbps. The maximum length of a 10BaseFL segment is 2000 meters, in which all segments are point-to-point with a transceiver at each end of the segment. The transceiver is further attached to two fiber optic cables through ST connectors, also called BFOC/2.5 connectors. With 10BaseFL, you can attain full-duplex transmission because one fiber optic cable transmits data, and the other receives data. 10BaseFL is frequently used on campus LANs where internetworks are set up between two or more buildings.

10BaseFiberBackbone (10BaseFB): Transmits data over a synchronous signaling link at 10 Mbps. The maximum segment length of a 10BaseFB segment is 2000 meters. Additional repeaters can be used in an extended Ethernet system because this system uses a synchronous signaling protocol.

10BaseFiberPassive (10BaseFP): Transmits data over a fiber optic passive star cabling system at 10 Mbps. The maximum segment length of a 10BaseFP segment is 500 meters (1640 feet). Up to 33 nodes can be connected to a segment. The 10BaseFP star is also considered a passive device because it requires no power and is ideal for locations where no active power supply is available. This system functions as a passive hub; it receives optical signals from MAUs and uniformly redistributes them to other 10BaseFP transceivers, including to the device from which the transmission originated.

Table 4.6 lists the specifications for 10BaseF.

TABLE 4.6   10BaseF Specifications

Characteristic	Specification
Rate of transmission	10 Mbps
Cable	Two multi-mode fiber optic cables
Topology	Star
Segment length	2000 meters (6562 feet)
Maximum transceivers/segment	2
Connectors	ST connector (also known as BFOC/2.5)
Frame transmission rate	0.65c (195,000 km/sec or 121,000 miles/sec where c is the speed of light)

Fast Ethernet

Two types of Ethernet include Fast and Gigabit. *Fast Ethernet* is a data transmission mechanism in LANs that can transfer data packets at a rate of 100 Mbps. It is based on the 10BaseT standard. In the development phase of Fast Ethernet, two different groups worked for Fast Ethernet standards: IEEE 802.3u and IEEE 802.12.

Fast Ethernet has retained the CSMA/CD access mode of the traditional Ethernet. It can be implemented using any cabling system, such as UTP, Shielded Twisted Pair (STP), and fiber optic.

There are four Fast Ethernet standards:

- 100BaseTX
- 100BaseT4
- 100BaseFX
- 100BaseSX

100BaseTX: Uses two pairs of Category 5 data-grade twisted-pair wires with a maximum distance of 100 meters (328 feet) between a hub and a node. Table 4.7 lists specifications for 100BaseTX.

TABLE 4.7   100BaseTX Specifications

Characteristic	Specification
Rate of transmission	100 Mbps
Topology	Star
Transmission medium	Two pairs of 100-ohm Category 5 UTP cables
Cable segment length	100 meters or 328 feet (half-duplex or full-duplex)

100BaseT4: Utilizes four pairs of cables, Category 3 or better, with a maximum distance of 100 meters (328 feet) between a hub and a node. Table 4.8 lists specifications for 100BaseT4.

TABLE 4.8   100BaseT4 Specifications

Characteristic	Specification
Rate of transmission	100 Mbps
Topology	Star
Transmission medium	Four pairs of 100-ohm Category 3 cables
Cable segment length	100 meters or 328 feet (half-duplex)

100BaseFX: Uses an optical cable for distances up to 2 kilometers (6561 feet or 1.24 miles) and is used to connect hubs over long distances in a backbone configuration. Table 4.9 lists the specifications for 100BaseFX.

TABLE 4.9   100BaseFX Specifications

Characteristic	Specification
Rate of transmission	100 Mbps
Topology	Star
Transmission medium	Two optical fibers
Cable segment length	412 meters or 1351 feet (half-duplex) and 2000 meters or 6562 feet (full-duplex)

100BaseSX: A proposed standard for Fast Ethernet over a fiber-optic cable using 850-nm wavelength optics. It is also called the Short Wavelength Fast Ethernet.

Figure 4.7 depicts the Fast Ethernet data transmission mechanism in LANs.

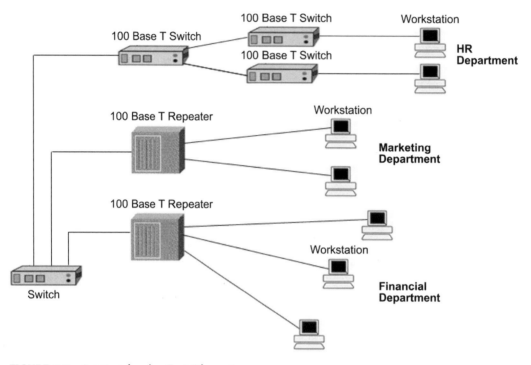

FIGURE 4.7   A network using Fast Ethernet.

Gigabit Ethernet (802.3z)

Gigabit Ethernet is a 1 Gbps (1000 Mbps) extension of the IEEE 802.3 Ethernet networking standard. This data transmission mechanism is used for corporate LANs, campus networks, and service provider networks. It is used to combine existing 10 Mbps and 100 Mbps Ethernet networks. A major advantage of using Gigabit Ethernet is that you can install it as a backbone network while retaining the existing investment in Ethernet hubs, switches, and wiring plants.

Gigabit Ethernet Transmission Mode

Gigabit Ethernet retains the CSMA/CD access method and the frame format as in the case of Ethernet and Fast Ethernet. It also supports both full-duplex and half-duplex modes of data transmission.

The frame size in the Gigabit Ethernet is the same as Ethernet. Gigabit Ethernet uses a bigger slot size of 512 bytes because the transmission rate is faster as compared to the Ethernet and Fast Ethernet. Gigabit Ethernet is used as a backbone between Fast Ethernet and Ethernet. Therefore, the carrier event (the sensing of collision occurrence during data transfer) in the MAC sublayer is increased while retaining the minimum frame size. If the frame size is shorter than the 512-byte limit, it is padded with an extension symbol. This process is called *Carrier Extension*. The disadvantage of this technique is that sending large numbers of padded packets reduces the throughput. To overcome this shortcoming, Gigabit Ethernet has introduced *Packet Bursting*. In this technique, when a node has to send a large number of data packets, only the first packet is padded, and the remaining packets are sent back-to-back. These packets have a minimum inter-packet gap until the burst timer of 1500 bytes expires. This technique increases the throughput on transmission.

Physical Medium of Gigabit Ethernet

Gigabit Ethernet works on two specifications:

1000BaseT: UTP copper cable

1000BaseX: STP copper cable

The 1000Base designation is an IEEE shorthand identifier. The "1000" in the media-type designation indicates the transmission speed of 1000 Mbps. "Base" refers to baseband signaling, which means that only Ethernet signals are carried through the medium.

The 1000BaseT specification supports full-duplex transmission over 100 meters (328 feet) of four pairs of Category 5 balanced copper cables. It is based on Gigabit Ethernet Physical layer specifications. Table 4.10 lists specifications for 1000BaseT.

TABLE 4.10 1000BaseT Specifications

Characteristic	Specification
Rate of transmission	1000 Mbps
Cable	Four pairs of Category 5 balanced copper cables
Segment length	100 meters or 328 feet
Maximum transceivers/segment	2
Connector	8-Pin RJ-45 connector
Signal encoding	PAM5

The 1000BaseX specification supports full-duplex transmission over an STP cable. It is based on the Fast Ethernet Physical layer specifications with GMII as the interface between the MAC sublayer and the Physical layer. Functionally, it is similar to AUI (in 10 Mbps) and MII (in 100 Mbps). This specification sets three standards, as follows:

1000BaseLX: Supports single and multimode optical fiber cables and uses long-wavelength lasers. Table 4.11 lists specifications for 1000BaseLX.

TABLE 4.11   1000BaseLX Specifications

Characteristic	Specification
Rate of data transmission	1000 Mbps
Cable	Two 62.5/125 or 50/125 *MultiMode optical Fibers* (MMF)
Segment length	316 meters (1037 feet)–5000 meters (16,404 feet or 3.1 miles)
Maximum transceivers/segment	2
Connectors	Duplex SC connector
Signal encoding	8B/10B

1000BaseSX: Transfers short-wavelength laser light over optical fiber to transfer data. Table 4.12 lists specifications of 1000BaseSX.

TABLE 4.12   1000BaseSX Specifications

Characteristic	Specification
Rate of data transmission	1000 Mbps
Cable	Two 62.5/125 or 50/125 MultiMode optical Fibers (MMF)
Segment length	275 meters (902 feet)–550 meters (1804 feet)
Maximum transceivers/segment	2
Connector	Duplex SC connector
Signal encoding	8B/10B

1000BaseCX: Uses copper jumper cables called *twinax* for data transmission. Table 4.13 shows the specifications for 1000BaseSX.

TABLE 4.13   1000BaseSX Specifications

Characteristic	Specification
Rate of data transmission	1000 Mbps
Cable	Shielded copper jumper cable
Segment length	25 meters (82 feet)
Maximum tansceivers/segment	2
Connectors	9-Pin shielded D-subminiature connector, 8-Pin ANSI Fiber Channel Type 2 (HSSC) connector
Signal encoding	8B/10B

Token Ring

Token Ring was first developed by IBM in the 1970s as an alternative to the Ethernet LAN. However, the Ethernet is more popular than Token Ring. Later, IEEE standardized Token Ring in the IEEE 802.5 specifications.

In a Token Ring, all nodes are connected in a ring topology, although the ring is logical and not physical. The nodes are connected to the central server, also called a Multistation Access Unit (MAU). Its primary function is to transmit or redirect data to the destination node.

In a Token Ring network, data is transmitted in the form of tokens. A token is also referred to as a *frame*. When a node possesses a token, it means that it can transmit data. When a node wants to send data, it takes the token, appends the information, and sends it to the next node. Until this information reaches the destination node, no other node can transmit data because there is only one token on the network. As a result, there are no collisions in a Token Ring network. When the transmission is complete, another node can send its token. Networks based on the Token Ring technology are also called *token-passing* networks. Token Ring supports two types of frames, *token frames* and *data frames*. Token frames consist of the following fields:

Start delimiter: Alerts nodes about the availability of the token.

Access control byte: Contains the priority field, reservation field, token bit, and a monitor bit.

End delimiter: Indicates the end of the frame or token in a logical sequence. It also indicates the presence of erroneous frames.

Data Frames consist of these fields:

Start delimiter: Alerts the nodes of data arrival.

Access control byte: Contains the priority field, reservation field, token bit, and a monitor bit.

Frame control byte: Indicates whether the frame contains data or control information. In case of control information, it indicates the type of control information.

Destination and source address: Stores the destination address of both the recipient and source nodes. The size of this field is 6 bytes.

Data: Indicates the length of a field for which a node can hold the token.

Frame Check Sequence (FCS): Denotes a value calculated by the source node by mapping with the frame contents to be transmitted. When the frame reaches the destination node, it is recalculated to verify whether the sent frame is a correct or a damaged one. If the values are not the same, the frame is discarded, and the source node is notified to send the frame again.

End delimiter: Sends signals indicating the end of the token or the data/command frame in a logical sequence. It also indicates the presence of any erroneous frames.

Frame status: Contains an address-recognized indicator and a frame-copied indicator. It is a 1-byte field terminating a command/data frame.

Figure 4.8 depicts the data frame supported by Token Ring.

Unlike Ethernet, a Token Ring has an excellent priority system. It can decide which nodes should be given priority to access the network. The priority and reservation fields in the Token Frame implement the priority system.

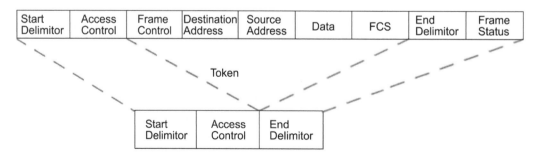

FIGURE 4.8 Format of the data frame supported by Token Ring.

The *fault management system* of Token Ring helps detect any problem, such as a broken cable or node failure. The system also helps correct network faults. Detection and correction are performed using the *Beaconing algorithm*, which detects, identifies, and repairs the problem areas in the network. For example, when a node detects a network fault it sends out a beacon frame indicating the failure domain. The failure domain includes the failed node and the next node in the network. When the identification is complete, beaconing begins the autoreconfiguration process wherein the failed node initiates the diagnostics to repair the fault.

Fiber Distributed Data Interface

Fiber Distributed Data Interface (FDDI) is a data transmission technology set by ISO. It uses fiber optic media in LANs to transmit information.

FDDI is based on the dual-ring architecture consisting of two rings—primary and secondary. The primary ring is used for data transmission, and the secondary ring is used as a backup in case of a primary ring failure. The transmission flow is also called *counter rotation* because the data on each ring flows in the opposite direction. The primary ring can transmit data at a rate of up to 100 Mbps whereas the secondary ring can support transmission rates up to 200 Mbps.

FDDI Fault Tolerance Features

The fault tolerance feature of FDDI enables networks to operate smoothly even when a failure occurs. Methods used for implementing fault tolerance in FDDI networks are as follows:

Dual-ring architecture: The architecture is based on two rings. The advantage of such architecture is that if one ring fails, the other ring takes over its function. As a result, data transmission is unaffected. For example, if a node on the dual ring stops functioning due to some technical error, this ring wraps into a single ring. This process is also called *dual wrapping*. The dual-ring architecture leads to enhanced network performance because there is no data loss.

Optical bypass switch: Bypasses traffic through an alternate route when a particular node fails. This ensures continuous data flow in the ring. The *optical bypass switch technique* prevents ring segmentation and removes nonfunctioning nodes from the ring.

Dual homing: Provides additional redundancy in a network. This technique is used in devices such as routers and mainframe hosts. For example, a router is connected to a pair of concentrators in which one concentrator link functions as the active link and the other remains passive for a backup. The passive link is activated only when the primary link fails, ensuring continuous data flow. For more information on concentrator links, see *http://www.cisco.com/univercd/cc/td/doc/cisintwk/ito_doc/fddi.htm*.

Transmission Media

FDDI uses optical fiber as the primary transmission media. It can also use copper cabling medium—Copper Distributed Data Interface (CDDI)—for data transmission. FDDI specifies two types of optical fibers:

Multimode fiber: Allows multiple modes of light to be transmitted across the fiber. Light enters the fiber from different angles and arrives at the end of the fiber at varied periods, resulting in the *modal dispersion* phenomenon. This phenomenon limits the bandwidth and distance covered. As a result, it is used in LANs to connect nodes in the same building.

Single-mode fiber: Allows only one mode of light to travel through the fiber. Modal dispersion does not occur in this fiber, resulting in higher bandwidth and better performance over longer distances. This is the reason a single-mode fiber is preferred for networks spanning two or more buildings.

FDDI Specifications

FDDI specifications cover the physical medium, that is, the cabling and the mode of data transmission. The functions of Media Access Control (MAC), Physical Layer Protocol (PHY), Physical-Medium Dependent (PMD), and Station Management (SMT) are as follows:

MAC: Resolves issues pertaining to frame format, token handling at the source and the destination nodes, problems related to node addresses, CRC, and error-recovery mechanisms.

PHY: Handles issues related to encoding and decoding of data signals and clocking requirements.

PMD: Handles physical medium component characteristics, such as cabling, fiber optic links, power levels, optical components, and connectors.

SMT: Handles the ring configuration; deals with ring control issues such as initialization, insertion, and removal of nodes in an FDDI network. SMT includes specifications for FDDI node configurations and ring (primary and secondary) configurations. SMT helps in fault isolation by determining the point of failure. When a failure is detected, SMT helps in recovery, scheduling, and statistics collection.

FDDI Frame Format

The FDDI frame format is similar to the frame format of Token Ring. It can support up to 4500 bytes of data (maximum size of the frame). The format of an FDDI frame includes:

Preamble: This is a unique sequence of bits which while transmitting data, notifies each node to receive the sent frame.

Start delimiter: Signals the nodes about the arrival of a frame.

Frame control: Contains the total size of the address fields. It also indicates whether the data is asynchronous or synchronous.

Destination address: This 6-byte field stores the address of the recipient node. This address can be a unicast, multicast, or a broadcast address.

Source address: This 6-byte field contains the address of the sender machine.

Data: Contains information sent by upper layers.

Frame Check Sequence (FCS): Calculates the CRC value at the source node. When the frame reaches the destination, a CRC is performed again to determine whether the sent frame has changed during transmission. If the calculated value is different, the frame is sent back or the source is notified to resend the frame.

End delimiter: Indicates the end of the frame using special symbols.

Frame status: Helps the source node to identify occurrence of an error during frame transmission.

FDDI Devices

The following are different types of FDDI connective devices that connect all the nodes in a LAN to the FDDI:

FDDI concentrator: Also called a *dual-attachment concentrator*. This concentrator connects primary and secondary rings, providing a backbone for the LAN. It ensures that network devices connected to FDDI do not collapse. Therefore, it enables smooth data transmission through the FDDI network.

Single-Attachment Station (SAS): Attached to the primary ring (unlike the concentrator). In turn, external devices are attached to the SAS. This ensures that the primary ring is unaffected in case of a single point failure of external devices.

Dual-Attachment Station (DAS): Has two ports, *A* and *B*. Each port connects DAS to the primary and secondary rings of the dual FDDI ring. The only disadvantage of connecting the DAS to the dual ring is that failure of DAS devices affects the functioning of the ring.

CISCO LAN EQUIPMENT

We have looked at the different media and hardware available for implementing LANs. Let us also look at the repeaters, switches, and routers manufactured by Cisco.

FastHub 400 Repeater

The Cisco FastHub 400 is a series of Fast Ethernet repeaters of 10/100 series that provide high performance, flexibility, and low connectivity cost. The 10/100 series repeaters also provide autosensing connectivity to the desktop. *Autosensing* enables each port to automatically configure the hardware to operate at the correct bandwidth.

A FastHub 400 repeater consists of managed and manageable 12- and 24-port 10/100 Fast Ethernet versions. In addition, it also consists of a manageable FastHub 412 and 424 series and FastHub 412M and 424M series. This 10/100 series provides 10/100 autosensing desktop connectivity at low cost. Figure 4.9 depicts an implementation of the FastHub 400 10/100 Series.

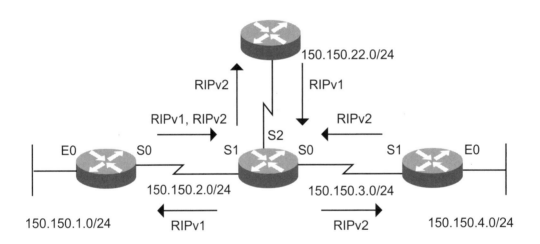

FIGURE 4.9   An implementation of a FastHub 400 10/100 Series hub.

Catalyst® 1900/2820 Series Switches

The Cisco Systems' Catalyst 1900 and 2820 series switches are available in Standard and Enterprise Editions, with models of varying port density, media type, and uplink options. The Catalyst 1900 series supports high-speed server and backbone connections using UTP wire or a combination of fiber optic and UTP. The Catalyst 2820 series switch has two expansion slots for high-speed interconnect modules—switched and shared 100BaseTX and 100BaseFX, ATM, and FDDI modules. Figure 4.10 depicts an implementation of the Catalyst 1900/2820 switches, placed on the network backbone.

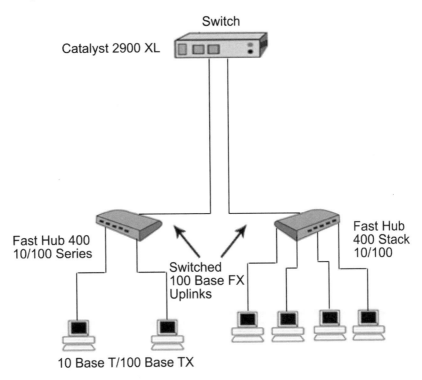

FIGURE 4.10   An implementation of a Catalyst 1900/2820 switch.

A Catalyst 1900 or 2820 series Standard Edition switch is effectively used in small or medium-sized business networks. These provide high-speed connectivity for individual desktops, 10BaseT hubs, servers, and other network resources. Catalyst 1900 or 2820 Enterprise Edition switches provide dedicated 10-Mbps connectivity to desktops and 10BaseT hubs. These switches connect users and workgroups to 100BaseT, 100BaseFX, ATM, or FDDI networks. In addition, they connect support port-based ISL or ATM VLANs across the network to other Cisco switches and routers. There are different models of Catalyst 1900 or 2820 switches. These models have different amounts of storage space for MAC addresses varying from 1 to 8 KB.

Table 4.14 lists the details and specifications of different models of the Standard Edition of Catalyst 1900 and 2820 switches.

TABLE 4.14   Models of Standard Edition Switches

Model	Specifications
WS-C1912-A	• Catalyst 1900 • 12 10BaseT & 2x 100BaseTX ports • 1024 MAC cache
WS-C1912C-A	• Catalyst 1900 • 24 switched 10BaseT ports • 1x 100BaseTX & 1x 100Base-FX ports • 1024 MAC cache
WS-C1924-A	• Catalyst 1900 • 24 10BaseT & 2x 100BaseTX ports • 1024 MAC cache
WS-C1924C-A	• Catalyst 1900 • 24 switched 10BaseT ports • 1x 100BaseTX & 1x 100Base-FX ports • 1024 MAC cache
WS-C2822-A	• Catalyst 2820 • 24 switched 10BaseT & 2x high-speed slots • 2048 MAC cache
WS-C2828-A	• Catalyst 2820 • 24 switched 10BaseT ports • 8192 MAC cache

Table 4.15 depicts specifications for Enterprise Edition switches.

TABLE 4.15   Models of Enterprise Edition Switches

Model	Specifications
WS-C1912-EN	• Catalyst 1900 • 12 10BaseT & 2x 100BaseTX ports • 1024 MAC cache
WS-C1912C-EN	• Catalyst 1900 • 24 switched 10BaseT ports • 1x 100BaseTX & 1x 100Base-FX ports • 1024 MAC cache
WS-C1924-EN	• Catalyst 1900 • 24 10BaseT & 2x 100BaseTX ports • 1024 MAC cache

TABLE 4.15   *(continued)*

Model	Specifications
WS-C1924C-EN	• Catalyst 1900 • 24 switched 10BaseT ports • 1x 100BaseTX & 1x 100Base-FX ports • 1024 MAC cache
WS-C2822-EN	• Catalyst 2820 • 24 switched 10BaseT ports • 2x high-speed slots • 2048 MAC cache
WS-C2828-EN	• Catalyst 2820 • 24 switched 10BaseT ports • 2x high-speed slots • 8192 MAC cache

Table 4.16 lists the specifications for Catalyst 2820 Modules.

TABLE 4.16   Catalyst 2820 Module Specifications

Module	Specifications
WS-X2811	1-port 100Base-TX module
WS-X2818	8-port 100Base-TX module
WS-X2821	1-port 100Base-FX module
WS-X2824	4-port 100Base-FX module
WS-X2831	FDDI UTP SAS module
WS-X2841	FDDI fiber SAS module
WS-X2842	FDDI fiber DAS module
WS-X2851	ATM 155Mbps UTP module
WS-X2861	ATM 155Mbps multimedia fiber module

Catalyst 2900 Switches

The Cisco Catalyst 2900 series provides 10/100 ports with Gigabit Ethernet up-links for higher speeds. In addition, Cisco has introduced the Catalyst 2948G-L2 and Catalyst 2948G-L3 switches for Layer 2 and Layer 3 intelligent enterprise switching.

Catalyst 2948G is a Layer 2 Ethernet switch. It has 48 RJ-45 10/100 ports and 2 Gigabit Ethernet uplink ports. These ports are connected with the help of the Gigabit Interface Converter (GBIC) interfaces. The Catalyst 2948G is made up of 48 10/100 ports, which have autosensing and autoconfiguring features with two Gigabit Ethernet uplinks. These uplinks enable midrange backbone frame processing and are responsible for wire-speed, nonblocking, and Layer 3 performance.

This switch provides high enterprise performance, which reduces traffic congestion and improves the response time of the network. In addition, it also provides low desktop protection cost.

Catalyst 3000 Series Switches

The Catalyst 3000 series has stackable switching architecture. These switches are capable of VLAN switching and are incorporated with stackable software. This series consists of a number of different switches and modules. Table 4.17 depicts details of switches and modules of Catalyst 3000 series.

TABLE 4.17   Switches and Modules in the Catalyst 3000 Series

Device	Type	Composition
Catalyst 3000 Switch	Stackable	• 16 10BaseT Ethernet ports • One AUI port • Two slots
Catalyst 3100 Switch	Flexible	• 24 dedicated 10BaseT ports • One FlexSlot • 3011 WAN access module • One Catalyst 3000 module
Catalyst 3200 Switch	Modular Chassis	• Seven expansion slots • One FlexSlot
Catalyst Matrix	8-port cross-point matrix	Three to eight Catalyst 3000, Catalyst 3100, or Catalyst 3200 switches
Catalyst 3000 Series Expansion Modules	Expansion	• 1-port 100BaseTX • 2-port 100BaseTX with Inter-Switch Link (ISL) • 3-port 10BaseFL • 1-port 100BaseFX • 2-port 100BaseFX with ISL • 4-port 10BaseT • 3-port 10BaseZ • 2-port 100VG (fiber and UTP) • 1-port ATM fiber modules

Catalyst 3500 Series Switches

The Cisco Catalyst 500 Series XL includes the Catalyst 3508G XL switch for stack aggregation. This stack aggregation of Layer 2 Gigabit Ethernet of Catalyst 3508G XL is formed with eight GBIC-based ports. This switch is also capable of autosensing 10/100BaseT Ethernet interfaces.

The Catalyst 3550-12G offers 10 GBIC-based Gigabit Ethernet ports and 2 10/100/1000 ports. These ports provide high-speed aggregation with Layer 2–4 capabilities. Table 4.18 depicts specifications of various models of Catalyst 3500 series switches.

TABLE 4.18 Specifications of Catalyst 3500 Series Switches

Model	Specifications
WS-C3524-PWR-XL-EN	• Catalyst 3524-PWR-XL • Enterprise Edition
WS-C3508G-XL-EN	• Catalyst 3508G XL • Enterprise Edition

Catalyst 3550 Series Switches

The Catalyst 3550 series switches are used in medium-sized enterprise wiring closets as powerful access layer switches. In addition, they are used in medium-sized networks as a backbone switch for medium-sized networks. Table 4.19 lists specifications of various Catalyst 3550 series switches.

TABLE 4.19 Specifications of Catalyst 3550 Series Switches

Model	Specifications
WS-C3550-12G	• 10 GBIC ports • 2-10/100/1000BaseT ports
WS-C3550-12T	• 10-10/100/1000BaseT ports • 2 GBIC ports
WS-C3550-24-EMI	• 24-10/100 and 2 GBIC ports • Enhanced Multilayer SW Image
WS-C3550-48-EMI	• 48-10/100 and 2 GBIC ports • Enhanced Multilayer SW Image
WS-C3550-24-SMI	• 24-10/100 and 2 GBIC ports • Standard Multilayer SW Image

(continued)

TABLE 4.19 *(continued)*

Model	Specifications
WS-C3550-48-SMI	• 48-10/100 and 2 GBIC ports • Standard Multilayer SW Image
WS-C3550-12G	• 10 GBIC ports • 2-10/100/1000BaseT ports
WS-C3550-12T	• 10-10/100/1000BaseT ports • 2 GBIC ports

Catalyst 3750 Series Switches

Cisco Catalyst 3750 Series Switches improve LAN operating efficiency because they are flexible and easy to use. This series of switches supports stacking of up to nine switches into a single logical unit for a total of 468 Ethernet 10/100 ports or 252 Ethernet 10/100/1000 ports. Each switch in the stack creates a 1:N (one standby for N active switches) availability scheme for Layer 2 and Layer 3 network controls.

The Cisco Catalyst 3750 switches are designed for deployment in similar topological positions in the network as the Cisco Catalyst 3550 switches. These positions include the access layer and the network backbone of enterprise wiring closets and branch office networks. The Cisco Catalyst 3750 switches are ideal for networks that require higher levels of availability, redundancy, and performance, and offer the next level of ease of stack management. Table 4.20 depicts specifications of various Catalyst 3750 series switches.

TABLE 4.20 Specifications of Catalyst 3750 Series Switches

Model	Specifications
WS-C3750-24TS-E	• Catalyst 3750 • 24 ports 10/100 Ethernet • 2 SFP Enhanced Multilayer Image
WS-C3750-24TS-S	• Catalyst 3750 • 24 ports 10/100 Ethernet • 2 SFP Standard Multilayer Image
WS-C3750-48TS-E	• Catalyst 3750 • 48 ports of 10/100 Ethernet • 4 SFP Enhanced Multilayer Image

TABLE 4.20 *(continued)*

Model	Specifications
WS-C3750-48TS-S	• Catalyst 3750 • 48 ports of 10/100 Ethernet • 4 SFP Standard Multilayer Image
WS-C3750G-24TS-E	• Catalyst 3750 • 24 ports of 10/100/1000T Ethernet • 4 SFP Enhanced Multilayer
WS-C3750G-24T-E	• Catalyst 3750 • 24 ports of 10/100/1000T Ethernet • Enhanced Multilayer Image
WS-C3750G-24TS-S	• Catalyst 3750 • 24 ports of 10/100/1000T Ethernet • 4 SFP Standard Multilayer
WS-C3750G-24T-S	• Catalyst 3750 • 24 ports of 10/100/1000T Ethernet • Standard Multilayer Image

Catalyst 4000 Series Switches

The Cisco Catalyst 4000 Series are modular switches, which include the Catalyst 4003 and the Catalyst 4006 chassis. The Catalyst 4003 is a modular 3-slot chassis. It consists of a switch supervisor engine in one slot and two open slots for switch modules. In addition to 24 Gbps of switching bandwidth, this switch provides expansion to 96 ports of 10/100 Ethernet or 36 ports of Gigabit Ethernet.

The Catalyst 4006 is a 6-slot chassis-based switch in which the first slot has a Supervisor III Engine. This 6-slot switch supports up to five interface modules. It is targeted at wiring closets, small backbones, and Layer 3 distribution points. The Supervisor III Engine implements a 64-Gbps switching fabric with integrated support for Layer 2, Layer 3, and Layer 4 switching.

Catalyst 4500 Series Switches

The Catalyst 4500 Series offers flexible Layer 2–Layer 4 switching and enhancing the control of converged networks. The Catalyst 4500 series is a next generation Cisco Catalyst 4000 Family platform and includes three new Catalyst chassis:

- Catalyst 4507R (7-slot: redundant Supervisor IV capable)
- Catalyst 4506 (6-slot)
- Catalyst 4503 (3-slot)

The Cisco Catalyst 4500 Series offers three chassis options and four supervisor engine options. It provides a common architecture that scales up to 240 ports. The Cisco Catalyst 4507R is the only Cisco Catalyst 4500 Series Switch to support 1 + 1 redundant supervisor engines with a subminute failover time. Because the Catalyst 4500 series is compatible with the Catalyst 4000 Series line cards and supervisor engines, the 4500 series can be deployed in converged networks (voice, video, and data).

Table 4.21 lists different features of the three series of Catalyst 4500 Chassis.

TABLE 4.21   Features of Cisco Catalyst 4500 Series Chassis

Feature	Cisco Catalyst 4503 Chassis	Cisco Catalyst 4506 Chassis	Cisco Catalyst 4507R Chassis
Total number of slots	3	6	7
Supervisor engine slots	1[1]	1[1]	2[2]
Supervisor engine redundancy	No	No	Yes
Supervisor engines supported	Supervisor Engine II Supervisor Engine II-Plus Supervisor Engine III Supervisor Engine IV	Supervisor Engine II Supervisor Engine II-Plus Supervisor Engine III Supervisor Engine IV	Supervisor Engine II-Plus Supervisor Engine IV
Line card slots	2	5	52

Catalyst 6500 Series Switches

The Cisco Catalyst 6500 Series offers up to 576 x 10/100/1000 Mbps Ethernet ports, high throughput multiple Gigabits, 10 Gigabit per second trunks, and scalable multilayer switching for both enterprise and service-provider networks.

The 6500 series optimizes IT infrastructure utilization that supports a number of services including data and voice integration and LAN/WAN/MAN convergence. The Cisco 6500 series switches include:

- Cisco WS-C6503 (3-slot chassis)
- Cisco WS-C6506 (6-slot chassis)

- Cisco WS-C6509 (9-slot chassis)
- Cisco WS-C6513 (13-slot chassis)

The Cisco Catalyst 6500 Series Switch is an intelligent multilayer modular switch that is designed to deliver converged services from the wiring closet to the core and from the data center to the WAN edge.

Cisco 1600 Series Routers

The Cisco 1600 Series Routers are ideal for linking small to medium-sized remote Ethernet LANs to regional and central offices over multiple WAN connections. These routers connect Ethernet LANs to WANs via ISDN, asynchronous serial and synchronous serial connections, supporting Frame Relay, leased lines, Switched 56, Switched Multimegabit Data Service (SMDS), and X.25.

The 1600 Series also provides support for ISDN connections and Internet links for customers or business partners to access the network without compromising the internal LAN traffic. Table 4.22 lists details of the Cisco 1600 Series models:

TABLE 4.22 Cisco 1600 Series Models

Model	Specifications
Cisco 1601 R	• One Ethernet • One Serial • One WAN interface
Cisco 1602 R	• One Ethernet • One Serial with integrated 56-Kbps DSU/CSU • One WAN interface card slot
Cisco 1603 R	• One Ethernet • One ISDN Basic Rate Interface (BRI) (S/T interface) • One WAN interface card slot
Cisco 1604 R	• One Ethernet • One ISDN BRI with integrated NT1 (U interface) • One S-bus port for ISDN phones • One WAN interface card slot
Cisco 1605 R	• Two Ethernet slots • One WAN interface card slot

Cisco 1700 Series Modular Access Routers

The Cisco 1700 Series are modular access multifunctional routers that provide routing, firewall, and VPN capabilities. These modular access routers have high

VPN performance up to full-duplex T1/E1 speeds and up to 100 VPN tunnels. These routers include hardware-based encryption and line speed performance features, and as a result, serve as a cost-effective integrated e-business platform for small and medium-sized businesses.

The Cisco 1700 Series routers support analog and digital communications in voice applications. They also provide easy Voice-over-IP (VoIP) and IP telephony deployments.

Cisco 2500 Series Routers

The Cisco 2500 series of Ethernet and Token Ring routers provide a variety of models designed for branch office and remote site environments. These routers provide a wide range of branch office solutions including integrated router/hub and router/access server models. Cisco 2500 routers have fixed configuration with at least two of these interfaces:

- Ethernet (AUI)
- 10BaseT Ethernet hub
- Token Ring
- Synchronous serial
- Asynchronous serial
- ISDN BRI

Each router can accommodate up to three WAN modules—two synchronous serial and one ISDN. The Cisco 2500 routers are based on the Flash EPROM technology for simplified software maintenance. These systems support a variety of Cisco IOS software features, such as IP, APPN, and RMON. Some of the models included in this series range from 2501 to 2516. In addition, these models include 2520, 2521, 2522, 2523, 2524, and 2525.

Cisco 2600 Series Routers

The Cisco 2600 series is a family of modular multiservice access routers that provide flexible LAN and WAN configurations, multiple security options, voice/data integration, and a range of high performance processors. The Cisco 2600 series routers share modular interfaces with Cisco 1600, 1700, 3600, and 3700 series routers. Recent additions to the Cisco 2600 series family of modular routers include the Cisco 2600XM models and Cisco 2691. These new models deliver extended performance, higher density, enhanced security performance, and increased concurrent application support. Some of the models in this series are 2610/11, 2610XM/11XM, 2612, and 2620/21. In addition, they include 2620XM/21XM, 2650/51, 2650XM/51XM, and 2691.

Cisco 3600 Series Routers

The Cisco 3600 series is a multifunctional platform that combines dial access, routing, and LAN-to-LAN services and multiservice integration of voice, video, and data in the same device. This series includes the Cisco 3660, the Cisco 3640, and Cisco 3620 multiservice platforms. As modular solutions, the Cisco 3660, the Cisco 3640, and Cisco 3620 routers have the flexibility to meet both current and future connectivity requirements. The Cisco 3600 series is fully supported by the Cisco IOS software, which provides features such as analog and digital voice capability, ATM access with T1/E1 IMA or OC-3 interfaces, and dial-up connectivity. Additional features include LAN-to-LAN routing, data and access security, WAN optimization, and multimedia features.

Table 4.23 depicts the series, models, and unique features of routers manufactured by Cisco.

TABLE 4.23   Cisco Router Series

Series	Models	Unique Feature
Cisco 1000 Series Routers	• Cisco 1005 • Cisco 1003	Suitable for ISDN or serial connections (Frame Relay, leased lines, X.25 or asynchronous dial-up) connections.
Cisco 10000 Series Routers	• Cisco 10008 • Cisco 10005	Provides Broadband Aggregation and Low-speed Private Line Aggregation Services (ATM/Frame Relay/Leased Line).
Cisco 10700 Series Routers	• Cisco 10720	Provides optimized optical transport with Dynamic Packet Transport (DPT), Cisco's Resilient Packet Ring (RPR) technology, to fully integrate IP routing and services.
Cisco 12000 Series Routers	• Cisco 12416 • Cisco 12410 • Cisco 12406 • Cisco 12404 • Cisco 12016 • Cisco 12012	Provides IP/MPLS, Peering, High-speed Private Line Aggregation, and Trunking services (ATM or Frame Relay transport over IP/MPLS).
Cisco 1600 Series Routers		Provides connections via ISDN, asynchronous serial and synchronous LAN-to-WAN connectivity via serial connections. Also supports Frame Relay, leased lines, Switched Multimegabit Data Service (SMDS), and X.25.

(continued)

TABLE 4.23 *(continued)*

Series	Models	Unique Feature
Cisco 1700 Series Modular Access Routers	• Cisco 1760 • Cisco 1751 • Cisco 1721 • Cisco 1720 • Cisco 1712 • Cisco 1711 • Cisco 1710 • Cisco 1701	Supports high-speed broadband and leased-line access, stateful firewall protection and intrusion detection, and multiservice data/voice integration applications.
Cisco 2500 Series Routers	• Cisco 2511 • Cisco 2509	This is a series of Ethernet and Token Ring routers providing branch office solutions including integrated router/hub and router/access server models.
Cisco 2600 Series Routers	• Cisco 2691 • Cisco 2651 • Cisco 2650 • Cisco 2621 • Cisco 2620 • Cisco 2613 • Cisco 2612 • Cisco 2611 • Cisco 2610	Provides flexible LAN and WAN configurations, multiservice voice/data integration, VPN access, Inter-VLAN routing, and Internet/intranet access with firewall security.
Cisco 3200 Series Mobile Access Routers	• Cisco 3251 • Cisco 3220	Provides secure data, voice and video communications, seamless mobility and interoperability across multiple wireless networks, ideal for vehicle integration.
Cisco 3600 Series Multiservice Platforms	• Cisco 3662 • Cisco 3661 • Cisco 3660 • Cisco 3640 • Cisco 3620	Provides solutions for data, voice video, hybrid dial access, Virtual Private Networks (VPNs), and multiprotocol data routing.
Cisco 3700 Series Multiservice Access Routers	• Cisco 3745 • Cisco 3725	Offers a single, integrated platform to combine flexible routing, low-density switching, office IP Telephony and Voice Gateway features. Also delivers internal inline power for the Ether Switch ports.
Cisco 6400 Series Broadband Aggregator		Offers Cisco IOS software, ATM switching and routing capabilities, and delivery of broadband network services, VPNs and voice services.

TABLE 4.23   *(continued)*

Series	Models	Unique Feature
Cisco 7200 Series Routers	• Cisco 7206 • Cisco 7204	Provides support for Fast Ethernet, Gigabit Ethernet, and Packet Over Sonet, Cisco IOS-based IP/MPLS feature support (QoS, Broadband Aggregation, Security, Multiservice, MPLS), multiprotocol support, full L2TP, PPP termination support, and scalability and flexibility; ideal for network redeployment.
Cisco 7300 Series Routers	• Cisco 7304 • Cisco 7301	Optimized for flexible, feature-rich IP/MPLS services at the network edge, where service providers and enterprises link together; ideal for intelligent, multigigabit network edge applications.
Cisco 7400 Series Routers	• Cisco 7401	These are compact, single-rack unit (RU) routers ideal for application-specific routing deployments in service provider and enterprise networks.
Cisco 7500 Series Routers	• Cisco 7513 • Cisco 7507 • Cisco 7505	Provides packet switching, distributed IP network services, including access control, QoS, and traffic accounting (NetFlow).
Cisco 7600 Series Routers	• Cisco 7613 • Cisco 7609 • Cisco 7606 • Cisco 7603	Provides high performance IP/MPLS features—Metro Ethernet Aggregation, Private Line Aggregation, Ethernet Subscriber Aggregation (PPPoE/LNS/SSG), WAN Aggregation and Headquarters Core Routing; ideal for service provider edge and enterprise MAN/WAN applications.
Cisco 800 Series Routers	• Cisco 837 • Cisco 836 • Cisco 831 • Cisco 828 • Cisco 827 • Cisco 826 • Cisco 813 • Cisco 811 • Cisco 806 • Cisco 805 • Cisco 804	Provides enhanced security for ISDN, serial connections (Frame Relay, leased lines, X.25, or asynchronous dial-up), IDSL, and ADSL connections; ideal for connecting small offices and telecommuters to the Internet or to the corporate LAN via ISDN, serial connections, IDSL, and ADSL.

(continued)

TABLE 4.23   *(continued)*

Series	Models	Unique Feature
	• Cisco 803 • Cisco 802 • Cisco 801	
Cisco MWR 1900 Mobile Wireless Router		Delivers a powerful solution that enables new applications and services in the Radio Access Network (RAN), reduces operating expenditures and the total cost of ownership of the RAN infrastructure; ideal for mobile wireless operators extending IP connectivity to the cell site edge of the RAN.
Cisco SOHO 70 Series Routers	• Cisco SOHO 78 G.SHDSL • Cisco SOHO 77 ADSL • Cisco SOHO 77 H ADSL • Cisco SOHO 76 ADSL • Cisco SOHO 71 Broadband	Provides affordable, multiuser access with a single DSL, easy setup, and deployment with a Web-based configuration tool; ideal for small office and home office customers.
Cisco SOHO 90 Series Secure Broadband Routers	• Cisco SOHO 97 • Cisco SOHO 96 • Cisco SOHO 91	Provides broadband access with integrated security features of Cisco IOS, including stateful-inspection firewall protection, VPNs, easy setup with Web-based setup tool, and advanced management capabilities.

Table 4.24 depicts series, models, and unique features of the switches manufactured by Cisco.

TABLE 4.24   Cisco Switch Series

Series	Models	Unique Feature
Cisco 6000 Series IP DSL Switches	• Cisco 6260 • Cisco 6160 • Cisco 6015	Provides support to multiservice applications with comprehensive IP+ATM managed services and common multi-DSL technologies for central offices (COs), remote terminals, and multiple dwelling units.

TABLE 4.24 *(continued)*

Series	Models	Unique Feature
Cisco BPX 8600 Series Switches	• Cisco BPX 8680 • Cisco BPX 8650 • Cisco BPX 8620	Provides scalable solutions to cost-effectively deliver ATM, Frame Relay, voice, circuit emulation, and IP services in medium sized PoPs or small organizations. Cisco BPX 8620 is used for broadband ATM services, Cisco BPX 8650 is used for broadband and MPLS, and Cisco BPX 8680 is used for broadband, narrowband, and MPLS services.
Cisco Catalyst 2900 Series Switches		Comprises industry's highest density, small form factor fixed configuration switches; offering feature-rich end-to-end software and solutions for workgroup and wiring closets.
Cisco Catalyst 2900 Series XL Switches	• Cisco Catalyst 2924 M XL DC	Comprises Software release SA3. Support 64 intraswitch VLANs and use virtual maps to enable real-time management.
Cisco Catalyst 2940 Series Switches	• Cisco Catalyst 2940-8TF • Cisco Catalyst 2440-8TT	Consists of Web-based Express setup, monitoring, and configuration tool and eight Fast Ethernet ports and one integrated Gigabit Ethernet port. Use outside wiring closet.
Cisco Catalyst 2950 LRE Series Switches	• Cisco Catalyst 2950 LRE 997 • Cisco Catalyst 2950 24 LRE • Cisco Catalyst 2950 8 LRE	Enables high availability and enhanced security, provides Gigabit Ethernet connectivity through fixed 10/100/1000 ports, embedded with Cluster Management Suite (CMS) Software.
Cisco Catalyst 2950 Series Switches	• Cisco Catalyst 2955C 12 • Cisco Catalyst 2955S 12 • Cisco Catalyst 2955T 12 • Cisco Catalyst 2950 24	Comprises fixed-configuration, stackable. and stand-alone switches and provides enhanced security, Quality of Service (QoS), and high availability features. It is incorporated with two sets of software.

(continued)

TABLE 4.24 *(continued)*

Series	Models	Unique Feature
	• Cisco Catalyst 2950 12 • Cisco Catalyst 2950C 24 • Cisco Catalyst 2950G 48 EI	
Cisco Catalyst 2970 Series Switches	• Cisco Catalyst 2970G-24T • Cisco Catalyst 2970G-24TS	Provides high wire-speed intelligent services and scale above 100 Mbps. They are Gigabit Ethernet switches providing high network performance and are embedded with Cisco Enhanced Image (EI) IOS Software.
Cisco Catalyst 3000 Series Switches		Enables Layer 2 and Layer 3 functions because of its stackable switching architecture. It comprises automatic error packet detection and elimination technology. Its features include low-latency switching, data integrity, and fault tolerance.
Cisco Catalyst 3500 Series XL Switches	• Cisco catalyst 3524 PWR XL • Cisco catalyst 3508G XL	Comprises stackable 10/100 and Gigabit Ethernet switches using Cisco Switch Clustering technology and GigaStack® GBICs. It allows single IP address switch management that enables high performance, manageability, and flexibility.
Cisco Catalyst 3550 Series Switches	• Cisco Catalyst 3550 48 EMI Switch • Cisco Catalyst 3550 48 SMI Switch • Cisco Catalyst 3550 24 DC SMI Switch • Cisco Catalyst 3550 24 EMI Switch	Provides high availability, Quality of Service (QoS), and security to enhance network operations. It is a stackable, multilayered switch with a range of Fast Ethernet and Gigabit Ethernet configurations.

TABLE 4.24 *(continued)*

Series	Models	Unique Feature
	• Cisco Catalyst 3550 24 FX SMI Switch • Cisco Catalyst 3550 24 PWR Switch • Cisco Catalyst 3550 24 SMI Switch	
Cisco Catalyst 3750 Series Switches	• Cisco Catalyst 3750-48TS Switch • Cisco Catalyst 3750-24TS Switch • Cisco Catalyst 3750G-24T Switch • Cisco Catalyst 3750G-24TS Switch • Cisco Catalyst 3750G-12S Switch	Features Cisco StackWise technology, a stacking architecture that offers exceptional levels of resiliency, automation, and performance. Customers can create a single 32-Gbps switching unit with up to nine individual Catalyst 3750 switches. Improves LAN efficiency.
Cisco Catalyst 4000 Series Switches	• Cisco Catalyst 4912G • Cisco Catalyst 4006 • Cisco Catalyst 4003	Delivers integrated resiliency and non-blocking, extending control from the backbone to the network edge with intelligent network services including advanced QoS, scalable performance, comprehensive security, and simple manageability.
Cisco Catalyst 4500 Series Switches	• Cisco Catalyst 4507 • Cisco Catalyst 4506 • Cisco Catalyst 4503	Offers wire-speed, multilayered switching with integrated resiliency, control, and intelligent services, including granular QoS, integrated inline power for IP telephony, software-based fault tolerance, and comprehensive management for converged network deployments.

SUMMARY

In this chapter you learned about the types, features, functions and specifications for provisioning hardware and media for LANs. You also looked at the various models and specifications of the switches and routers manufactured by Cisco. In the next chapter, we will move on to provisioning hardware and media for WANs.

POINTS TO REMEMBER

- Repeaters receive signals from a segment, and then amplify and retransmit them to the other segments of the network.
- Repeaters are used when you need to extend your network beyond 100 meters (328 feet). However, you can use only five repeaters in a series to extend the cable length for a single network.
- The 5-4-3 rule states that there can be only a maximum of five segments, connected through four repeaters/concentrators between two communicating nodes.
- Passive hubs have one 10Base2 port connected to each LAN device through RJ-45 connectors.
- An active hub repairs damaged data packets and retimes the distribution of other packets.
- Intelligent hubs transmit data at the rate of 10, 16, and 100 Mbps to desktop systems using standard topologies such as Ethernet, Token Ring, or FDDI.
- Routers segment a network, limit broadcast traffic, and provide security, control, and redundancy between individual broadcast domains.
- Ethernet is a network technology used for transmitting data packets between nodes in a network. It uses the CSMA/CD system for data transmission.
- The Media Access Control (MAC) sublayer is responsible for transmitting information through an Ethernet network.
- The elements of an Ethernet frame are PRE, SOF, DA, SA, Length/Type, MAC Client Data, and FCS.
- The half-duplex mode of data transmission means that data cannot be transmitted and received at the same time.
- The full-duplex mode of data transmission allows simultaneous transfer of data between two nodes.
- Fast Ethernet is a data transmission mechanism in LANs that can transfer data packets at the rate of 100 Mbps.
- Gigabit Ethernet is a 1-Gbps (1000 Mbps) extension of the IEEE 802.3 Ethernet networking standard.

- In a Token Ring, all nodes are connected in a ring topology. However, the ring is a logical, not a physical structure.
- FDDI is based on the dual ring architecture consisting of two rings, primary and secondary.
- Dual homing is a fault tolerance technique that provides additional redundancy in the network.
- FastHub 400 repeaters, Cisco series switches, and Cisco series routers are used for designing LANs.

5 Media and Hardware Selection for Provisioning WAN

IN THIS CHAPTER

- WAN Design Considerations
- Provisioning Hardware for WANs
- Provisioning Media for WANs
- WAN Technologies
- WAN Protocols

In today's high technology scenario, it is impossible to have all resources, people, or equipment at a particular area. The need of the hour is a well-designed and connected network. If the resources are spread over more than one country, we need a WAN setup. For example, if an organization's head office is located in the U.S., and it has branch offices in the U.S. as well as other countries, then all the offices need to be connected through a network for information sharing. This is possible through a robust and well-organized WAN.

A WAN setup includes obtaining media and hardware to build up a cost-effective, high-bandwidth WAN. An important step towards designing a WAN is selecting the appropriate hardware and media. An incorrect assessment of the hardware and media required for the proposed network can result in poor network performance and significant financial losses. To select the correct hardware and media, you need to have an in-depth knowledge of the types, features, functions, and specifications of the different hardware and media used for designing a WAN.

WAN DESIGN CONSIDERATIONS

The design considerations for a WAN are different from a LAN, as a WAN involves setting up network infrastructure on a much wider scale with greater numbers of users on the network as compared to a LAN. A WAN user can be a single user using dial-up services to get connected to the Internet or can be a part of a big organization that is spread globally. In both these scenarios there are certain considerations that must be kept in mind while setting up a WAN network infrastructure. A WAN network should be designed to provide the following to its end users:

- Efficient bandwidth utilization
- Minimum cost
- Maximum service

While designing WANs, some important design considerations are as follows:

Backbone load determination: Accessing network traffic load consists of a detailed survey of users' physical locations relative to the destinations, identifying the applications sending traffic through the network, analyzing whether the applications are bandwidth-hungry such as multimedia applications, and calculating the quantity of data to be transmitted through the proposed network. This information gives a clear picture of the client requirements and helps in shaping the WAN topology.

Bandwidth distribution: The proposed WAN design should be such that it meets the bandwidth requirements of the user terminals connected in the network.

Distance limitation calculations: The WAN infrastructure should be designed in such a way that the distance between the terminals should not slow down the data transmission rate within the network.

Unlike LANs, WANs are global networks catering to a far wider user base; as such, WAN uses different types of devices depending upon the environment.

PROVISIONING HARDWARE FOR WANs

The types of WAN devices discussed in this section are:

- WAN Routers
- WAN Switches
- WAN Access Servers
- Modems
- Terminal Adapters
- CSU/DSU
- Data Terminal Equipment (DTE)
- Data Communications Equipment (DCE)

WAN Routers

Routers used in WANs perform the same function as in LANs. For example, in LANs, routers are used to connect two or more network segments; in WANs, they are used to connect two or more LANs or campus LANs. They act as the junction points of the WAN networks. In addition, routers in a WAN also help in determining the route for data transmission. Cisco 2600 series, Cisco 3600, and Cisco 3810 series are WAN routers.

WAN Switches

In a WAN environment, switches perform the same functions as in a LAN. Switches are Data-link layer devices that help to interconnect and relay information in multiple LAN segments in a WAN. In simple words, all the WAN links are connected together with WAN switches. In WANs, switches function as multiport devices used in carrier networks. WAN switches are used to transmit data traffic of Frame Relay and X.25. Cisco Series IGX 8410, 8420, BPX 8620, and 8650 are WAN switches. Figure 5.1 illustrates a WAN switch.

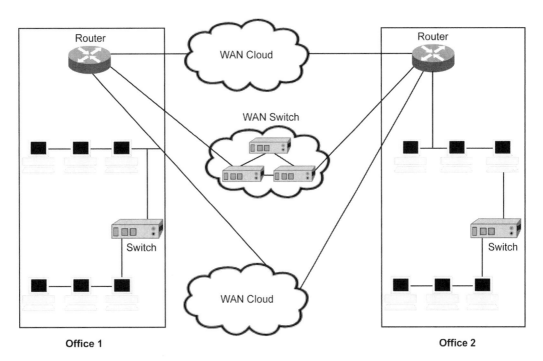

FIGURE 5.1 A WAN switch connecting different WAN links.

WAN Access Server

An access server is a device or server that handles all dial-in and dial-out services from the network users by acting as a connecting point. This helps the remote users situated anywhere geographically to access the network. The functions of an access server are:

Authenticating: When the users connect to the access server using a modem, an access server authenticates the users by checking for valid usernames and passwords. After they are validated, the server authorizes them to use the network resources.

Authorizing: Access servers authorize authenticated network users to use network resources. For example, when the users using dial-up connections are connected to the access server, they can start sending requests for a Web page or image file.

Accounting: Access servers take into account the logging or the usage time of the users. This is useful in case of private WANs where billing and user accounts have to be taken care of.

Figure 5.2 shows WAN access servers.

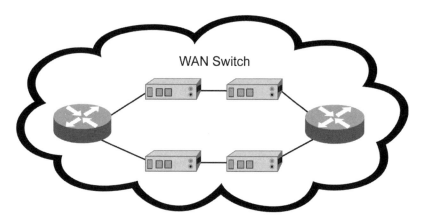

FIGURE 5.2   A WAN access server connecting different users.

Modems

Modems are electronic devices that are used for converting digital signals into analog signals and vice versa for successful transmission of data across the network. These are of two types:

Analog modem: For transmitting data, a modem situated at the sender's end interprets the digital signals and converts them into analog signals so that they can be transported across cable lines or telephone lines. Another modem situated at the receiver's end reconverts the analog signals back into digital signals. Figure 5.3 depicts a modem.

Cable modem: Cable modems are similar to analog modems, but offer high transmission rates ranging up to 1.5 Mbps as compared to the analog modem (53 Kbps). It is one of the latest entries into the telecommunication industry and is used in large WANs to transfer data across the network. For transmitting data, a cable TV line (coaxial cable) is used to send and receive data across the network. A cable modem can serve as either an internal device (inside the computer) or as an external unit.

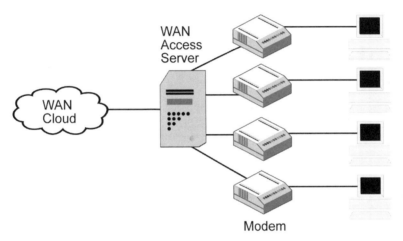

FIGURE 5.3 Modems converting and reconverting data in the form of signals.

Terminal Adapter

A *terminal adapter* is installed in the workstation and acts as an interface between a workstation and an ISDN line. It is also called an *ISDN terminal adapter*. A terminal adapter can be considered as a replacement for a modem in case of organizations using ISDN for connecting to the network. Unlike the modem, an ISDN terminal adapter does not have to convert the signal from analog to digital because it carries data in a digital form from the beginning. Figure 5.4 illustrates a terminal adapter.

CSU/DSU

Channel Service Unit/Data Service Unit (CSU/DSU) acts as an interface to connect a router to a digital circuit. The function of CSU/DSU is similar to an external modem;

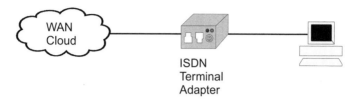

FIGURE 5.4 A terminal adapter providing ISDN connectivity.

it converts the digital frame used in LANs into a frame format appropriate to a WAN-specific environment. The CSU part of the CSU/DSU is responsible for receiving and transmitting signals from and to the WAN. Further, it protects the signals from electrical interference. The DSU takes care of timing errors and signal regeneration.

Data Terminal Equipment (DTE)

These devices serve either as the start point or end point of the transmitted data frames. Examples of DTE devices are PCs, print servers, and file servers.

Data Communications Equipment (DCE)

These devices are immediately responsible for receiving and forwarding the data packets from the start points to their end points. Examples of DCE devices are switches, routers, bridges, and repeaters.

WAN Interface Card

This refers to a computer circuit board or card installed at users' workstations in order to facilitate connectivity with the network. Table 5.1 lists various WAN devices and their models.

TABLE 5.1 WAN Devices and Models

Hardware	Series/Models
WAN Access servers	• AS2509/2509-ET • AS2509-RJ • AS2511 • AS2511-RJ • AS 2512
WAN Interface cards—Serial WIC	• Serial WIC • WIC-1T • WIC-2T • WIC-1DSU-T1

TABLE 5.1 *(continued)*

Hardware	Series/Models
	• ISDN BRI WIC
	• WIC-1B-S/T
	• WIC-1B-u
	• DSL WIC
	• WIC-1ADSL-1-DG
	• WIC-1-ADSL
	• WIC-1-SHDSL
	• Analog Modem WIC
	• WIC-1AM
	• WIC-2AM
	• T1/E1 & G.703 Multiplex Trunk Voice WIC
	• VWIC-1MFT-T1
	• VWIC-2MFT-T1
	• VWIC-2MFT-T1-D1
	• VWIC-1MFT-E1
	• VWIC-1MFT-G703
	• VWIC-2MFT-E1
	• VWIC-2MFT-E1-D1
	• VWIC-2MFT-G703
Cisco WAN routers	• Cisco 12000 series—used in building IP/MPLS networks.
	• Cisco 10700—used in delivering Ethernet and IP services in WAN.
	• Cisco 10000—used for delivering high-density T1 aggregation and advanced IP services.
WAN switches	• Cisco BPX 8600 series—basically an ATM switch. It provides switching ATM, multiservice adaptation, and aggregation.
	• Cisco IGX 8400—used to carry legacy data, and voice data over any media such as Frame Relay, ATM, and others.
	• Cisco MGX 8000—supports VoIP, VoATM, MPLS support, and multihosted softswitch support.

PROVISIONING MEDIA FOR WANs

As the name suggests, WANs are spread over a large area and may connect strategic locations over a country or more than one country. Depending upon the area covered, client's budget issues, traffic patterns, bandwidth requirements, and scalability, you can decide upon the type of WAN media to be selected for your client. For this, you need to grasp the specific needs of the users and applications that would be traversing the network. The most common choices of WAN media are standard telephone lines, digital lines, and leased lines, as they provide cost-effective services. Other WAN media offering high-cost services are optical fiber links and wireless technologies such as satellite and cellular transmission. Here we will discuss in detail the various media options available for provisioning WANs. In this section you will learn about various media options available in WAN, such as:

- Leased lines
- Network cloud
- Fiber links
- Wireless

Let us learn about each of them in detail.

Leased Lines

Also known as *private lines*, leased lines offer a point-to-point circuit between two endpoints (customer and the service provider in a WAN). A leased line provides an ideal solution for bandwidth-demanding low-latency multimedia applications such as voice and video. The leased line can be rented or hired from telecommunication organizations or vendors (carriers) for transferring data either on the Internet or any other WAN. Leased lines give the highest bandwidth, maximum control, and maximum use, but they are also very costly. The distance covered and the speed at which the data is transmitted by the leased line carrier are used to calculate the cost of the connection. This kind of service is used by business organizations that require high-speed and low-error data transmissions. They are best suited under long connect times and shorter distances.

Leased lines are of various types, such as:

- T/E carriers
- SONET
- DSL

T/E Carriers

These are of two types, namely T1/E1 and T3/E3.

T1/E1: *T1* is a dedicated phone line providing a data transmission rate of about 1.544 Mbps. It contains 24 channels, and each of them is capable of transmitting data at the rate of 64 Kbps. An organization buys only few channels as per the need. This kind of channel access is called *Fractional T1* access. While T1 is a standard used in North America, *E1* is more popular in Europe and Asia. E1 contains 30 channels, and each of them is capable of transmitting data at the rate of 2.048 Mbps. T1/E1 lines are also known as *DS1* lines.

T3/E3: *T3* is a dedicated phone line offering a data transmission rate of about 43 Mbps. It is used in case of high-speed WAN media requirements. It offers 672 individual channels, and each of them is capable of transmitting data at the rate of 64 Kbps. While T3 is a standard used in North America, *E3* is more popular in Europe and Asia. T3/E3 lines are also known as *DS3* lines.

Synchronous Optical Network (SONET)

SONET is used as a standard for synchronous mode of data transmission across an optical media. The international equivalent of SONET is *Synchronous Digital Hierarchy* (SDH). SONET has a basic data transfer rate of 51.84 Mbps and a set of multiples called *Optical Carrier levels* (OCx).

Table 5.2 lists the data transfer rates specified by various optical levels of SONET.

TABLE 5.2   Data Transfer Rates by SONET Standards

Optical Level	SDH Equivalent	Line Rate (Mbps)	Bandwidth availability (Mbps)
OC-1		51.840	50.112
OC-3	STM-1	155.520	150.336
OC-12	STM-4	622.080	601.344
OC-48	STM-16	2488.320	2405.376
OC-192	STM-64	9953.280	9621.504
OC-768	STM-256	39813.120	38486.016

DSL

Digital Subscriber Line (DSL) is a technology that uses sophisticated modulation schemes to transfer high-bandwidth digital data over ordinary copper lines. DSL is used in case of small business and home users and is used only for connections from a telephone switching station to a home or office and not between switching stations. DSL lines can carry both general and voice data. Different types of DSL (also referred to as xDSL) are:

- ADSL
- CDSL
- G. lite or DSL lite
- HDSL

Asymmetric Digital Subscriber Line (ADSL)

ADSL enables data transfer in two ways:

Downstream: Data is transferred from the ISP to the user.

Upstream: Data is transferred from the user to the ISP.

Using downstream ADSL, data can be transferred at the rate of 6.1 Mbps, and in case of upstream, the transfer rate is 640 Kbps.

Consumer DSL (CDSL)

CDSL offers a slower rate of data transfer as compared to ADSL. It gives the data transfer rate of 1 Mbps downstream and less than 640 Kbps upstream.

G.Lite or DSL Lite

G.Lite or *DSL Lite* is a slower version of ADSL. It transfers data at the rate of 1.544 Mbps to 6 Mbps in downstream and 128 Kbps to 384 Kbps in upstream.

High bit-rate Digital Subscriber Line (HDSL)

HDSL follows a symmetric mode of transmission, wherein an equal amount of data is transferred in both directions (upstream and downstream), unlike ADSL. It is used for wide-band digital transmission within an organization.

Components of Leased Lines

Leased lines consist of the following components:

Customer Premises Equipment (CPE) modems: Refers to modems installed at the client end to receive and send data across a WAN

Carrier-end modems: Refer to modems installed at the carrier (telecommunication organization selling leased lines) end

Transmission medium: Refers to the physical medium that actually connects the two endpoints in a WAN

Figure 5.5 illustrates a leased line setup.

Network Cloud

The *network cloud* is the space or area enclosed between two endpoints in a WAN. When a workstation sends data across the network, the data passes through series of

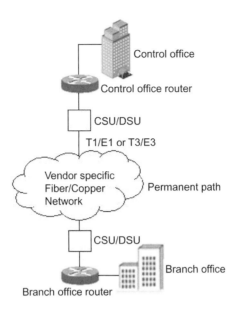

FIGURE 5.5   A leased line setup with its components.

equipment and protocols before reaching the destination. This space that includes the protocols, technologies, and equipment is known as a *cloud*. Network devices such as routers, switches, bridges, and multiplexers are all part of the network cloud.

Network data transmission speed can be optimized by arranging the network cloud in such a way that it gives maximum throughput to the customer.

NOTE

Fiber Links

Fiber links in a WAN are established on optical fibers. In case of fiber links, data is transmitted across the network through the optical fiber instead of the conventional copper cable. Data travels in the form of light pulses across glass or plastic fibers. In case of optical fiber transmission, repeaters are required at regular intervals.

There are two modes used in optical fiber data transmission:

Multimode optical fiber: Allows multiple modes of light to be transmitted across the fiber. This process of allowing multiple modes of light dampens the speed of transmission. As a result, this fiber is used for shorter distances, such as in LANs.

Single mode optic fiber: Allows only one mode of light to travel through the fiber, which means transmission is faster. It is used over long distances and in LANs spanning more than one building.

The major advantages of fiber links are faster data transmission, and greater support for multimedia applications such as picture-, video-, and voice-based applications Fiber links are used in MANs and are not useful for home users because the setup cost is very high.

Wireless

Using wireless media, data is transferred through high-frequency radio waves between two endpoints in a network. The data transmission rate in wireless media is very high as compared to other WAN media but due to high cost and low reliability, wireless is not often used. Wireless LANs (WLANs) are a good example of wireless media. WLAN components consist of fixed-position transceivers called base stations, which send a signal within a specified area called microcells. Each transceiver is then connected to other transceivers so that each user can communicate with the others in the network.

Other technologies using wireless media are: infrared, spread spectrum radio, frequency-hopping spread spectrum, and narrowband.

WAN TECHNOLOGIES

As communication in a WAN occurs over a geographically vast area, the data travels through longer distances and encounters more disturbances—physical as well as logical—in a WAN, as compared to a LAN. Therefore, we have a wide range of technologies with services and features offered that could be implemented per the client's needs and specifications. WAN technology is divided into the following two major categories:

- Circuit switching
- Packet switching

Circuit Switching

Circuit switching is a data transmission technology wherein a dedicated communication path or *channel* is set up between two endpoints or end workstations. This channel is also called a *circuit*. In a circuit-switching technology, separate setup messages are sent to establish a circuit. When the setup is established, data transmission starts. Circuit-switching services are also referred to as *Time Division Multiplexing* (TDM) services. In TDM, data is transferred in fixed time slots, which helps in optimizing bandwidth. Therefore, it is used in case of voice and video applications. The Public Switched Telephone Network (PSTN) and Integrated Services Digital Networks (ISDN) use circuit switching technology.

Packet Switching

As the name suggests, *packet switching* is a data transmission technology that uses packets as a means of transferring data across the network. The packets are routed to their destination using the destination addresses contained within the packets. This kind of communication between the sender and recipient is known as *connectionless communication*. Frame Relay and ATM are good examples of the packet-switching technology. The Internet is another example of a connectionless network that uses packet-switching as a means of data transmission.

Public Switched Telephone Network (PSTN)

PSTN uses circuit-switching technology for data transmission. It is also known as Plain Old Telephone Service (POTS). As the term suggests, POTS provides analog-voice dial-up services using the normal telephone line in a network. The medium used in sending the voice data is copper wire. In a network using POTS, a modem is connected to the sender's end and another is connected to the recipient's end. If the networks do not use individual modems per computer, then a common LAN modem is used to process users' requests in a particular LAN segment within a WAN. A POTS-based network is restricted to a data transmission rate of up to 53 Kbps. The bandwidth availability is also less as compared to other technologies used in WAN. Figure 5.6 shows a PSTN setup.

To overcome the problem of bandwidth and user expandability of PSTN, digital communications came into the picture. Currently, ISDN is considered a better option for WAN environments. The specifications promulgated by the International Telecommunications Union (ITU) enabled inexpensive ISDN infrastructure being offered with high-bandwidth Internet access, which resulted in ISDN gaining popularity in international markets.

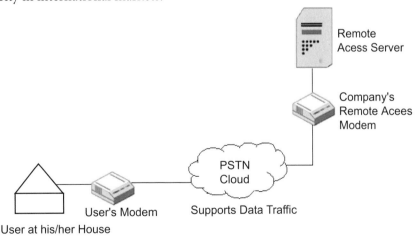

FIGURE 5.6 Data transfer in a network through a PSTN.

Integrated Services Digital Network (ISDN)

ISDN is an international standard for transmitting video, voice, and other data over digital as well as analog telephone wires. It uses circuit-switching methodology for data transfer between endpoints. In addition, it is faster in processing users' requests as compared to PSTN and offers a data transmission rate of 64 Kbps and above as compared to 53 Kbps in PSTN. ISDN involves digitization of the older telephone line infrastructure in order to transmit data involving applications using voice, graphics, music, and video. Figure 5.7 shows an ISDN setup.

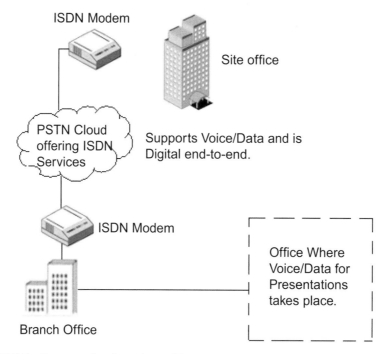

FIGURE 5.7 Data transfer through an ISDN setup.

Apart from the common network devices such as routers and bridges, there are certain devices specific to ISDN. Let us discuss these ISDN-specific devices in detail.

ISDN Devices

The various ISDN-specific devices are as follows:

■ Terminals
■ Terminal Adapters
■ Network terminal devices

Terminals

Terminals used in ISDN services can be categorized into the following:

- Specialized ISDN terminals, also referred to as *Terminal Equipment type 1* (TE1)
- Non-ISDN terminals, also referred to as *Terminal Equipment type 2* (TE2). Examples are DTE devices such as modems and fax machines.

The major difference between TE1 and TE2 is that while TE1 connects to the ISDN network through a four-twisted-pair digital link, TE2 connects to the ISDN network through Terminal Adapters (TA).

Terminal Adapters (TAs)

TAs are devices that act as an interface or connection point between non-ISDN equipment, such as analog phone lines/terminals.

Network Terminal Devices

These devices connect TE1/TE2 to the two-wire local loop. They are categorized as NT1 and NT2. NT1 is packaged with network components provided by the carrier. NT2 on the other hand is a specialized device found in Private Branch Exchanges (PBXs).

Along with devices, ISDN specifications also include reference points for logical interfaces between network equipment such as TAs and NT1s. Table 5.3 contains the reference points with their descriptions.

TABLE 5.3 Reference Points with Descriptions

Reference Point	Description
R	Used between non-ISDN devices and terminal adapters.
S	Used between user terminals and NT2 type equipment.
T	Used between NT1 and NT2.
U	Used between NT1 and NT2. It is valid only in the U.S.

Figure 5.8 shows the configuration of an ISDN setup.

There are two ISDN devices and one non-ISDN device (standard telephone). The ISDN equipment uses the "S" reference point, and the non-ISDN equipment uses "R" reference point.

ISDN Layer Specifications

The basic channel of information flow in an ISDN is a frame. The ISDN devices discussed here help in connecting the vast ISDN network and therefore assist in the

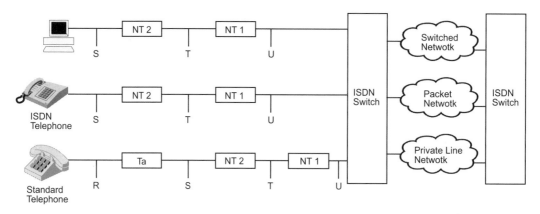

FIGURE 5.8 ISDN configuration setup connecting different devices.

information flow across the network. The implementation and functionality of ISDN spans over Layer 1, Layer 2, and Layer 3. Let us explore the functional aspects of ISDN with the help of these layer specifications.

ISDN Layer 1 Specifications

Layer 1 has a set structure for a frame format. The structure of a frame depends on whether the frame is an outbound frame (information flowing from the workstation to the network) or an inbound frame (information flowing from the network to the workstation).

The basic Layer 1 frame structure in case of both inbound and outbound frames is shown in Figure 5.9.

The functions of the main elements of the ISDN Layer 1 frame format are as follows:

F (framing bit): Responsible for synchronization

L (load balancing): Responsible for adjusting the average bit value

E (echo of previous D-bit): Responsible for channel contention resolution when many workstations are sending data

A (activation bit): Responsible for preparing devices for sending and receiving data

S (spare bit): Not yet assigned

B1, B2, and D (channel bits): Responsible for handling user data

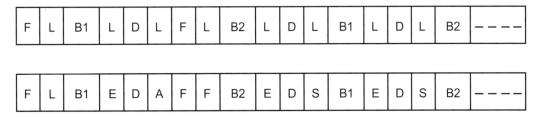

FIGURE 5.9 ISDN Layer 1 frame structure in inbound and outbound frames.

ISDN Layer 2 Specifications

In ISDN Layer 2 (Data-link layer), the data flows in the form of Link Access Procedure D channel (LAPD). "D channel" indicates the Delta channel that is responsible for controlling and signaling. Layer 2 handles all problems pertaining to controlling and signaling of information flow in the network.

The functions of main elements of the ISDN Layer 2 frame format are as follows:

Flag field: Helps in streamlining the beginnings and endings of streams.

Address field: Consists of various subdivisions:

Service Access Point Identifier (SAPI): Helps in identifying the portal at which LAPD services are provided in case of Layer 3.

Command or a Response (C/R): Helps in determining whether a frame has a command or response signal contained in it.

Extended Addressing (EA): Indicates whether the specified byte is the last addressing field. If the value is found to be 1, then the byte is determined to be the last byte in the field.

Terminal Endpoint Identifier (TEI): Helps in identifying the broadcasting workstation, that is, the workstation from which the signal has been broadcast.

Control: Handles the congestion control features in the network.

Data: Contains the actual data to be transmitted.

Frame Check Sequence (FCS): Handles error correction and verifies whether the frame has reached its destination intact.

ISDN Layer 3 Specifications

ISDN Layer 3 specifications are useful in ISDN signaling in which communication occurs between two or more network devices. Under this layer, two different standards specified are ITU-T I.450 and ITU-T I.451. These specifications help in streamlining circuit-switched and packet-switched connections and ensure safe transmission of information. In addition, these help in establishing initial contact between the two terminals. In case of traffic congestion, it sends a call termination signal to end the transmission. These functions are performed through series of standard messages such as SETUP, CONNECT, RELEASE, USER INFORMA-TION, CANCEL, STATUS, and DISCONNECT.

ISDN Services

ISDN offers three types of services. These are as follows:

- Basic Rate Interface (BRI)
- Primary Rate Interface (PRI)
- Broadband ISDN (B-ISDN)

Basic Rate Interface (BRI)

BRI services are used in small enterprise ISDN network setups and home setups. To carry data, BRI uses two channels, B-channel (B stands for *bearer*) and D-channel (D stands for *delta*). B-channel is associated with transmission of data—voice as well as video—and D-channel is responsible for controlling and signaling the transmitted data. A BRI consists of two 64 Kbps B-channels and one 16 Kbps D-channel.

Primary Rate Interface (PRI)

Primary Rate Interface is used in case of large network setups having many users. It contains 23 B-channels, one 64 Kbps D-channel on a T1 line (in the U.S., Canada, and Japan) and 30 B-channels and 1 D-channel in case of E1 lines (other countries).

Broadband ISDN (B-ISDN)

Broadband ISDN deals with multimedia services in the carrier backbone. In fact, B-ISDN is a precursor to the ATM technology, but with time, ATM has gained more popularity than B-ISDN. It has a transfer rate of up to 52 Mbps. SONET is the physical backbone of B-ISDN networks.

Frame Relay

Frame Relay is a technique used for data transmission between LANs and end-points of WANs. Frame Relay is an access protocol that functions between user devices, such as a LAN bridge and a network. It provides a unique ability for the organization's network to connect many remote sites across a single physical connection. This reduces the number of point-to-point physical connections that any other network would normally require, consequently making the whole setup cost effective.

NOTE

Initially, Frame Relay was designed as a protocol to be used with ISDN as a frame-switching component of the ISDN network, but currently, it is used with other network interfaces also and sold as an independent service.

Frame Relay was standardized by the International Telecommunication Union-Telecommunications Standards Section (ITU-T). Since then, its specifications are followed worldwide. In the U.S., the American National Standards Institute (ANSI) sets Frame Relay standards.

Both ITU-T and ANSI have given their standards for Frame Relay. Table 5.4 illustrates differences between these two standards.

TABLE 5.4   Difference between ITU-T and ANSI standards

ITU-T Standard	ANSI Standard	Description
I.233	T1.606	Sets the Frame Relay architecture and service
Q.922 Annex A	T1.618	Sets the Data-link layer core aspects.
Q.933 Annex A	T1.617 Annex D	PVC Management
I.370	T1.606a	Management of congestion
Q.933	T1.617	SVC Signaling
Q.920	T1.602	Description of LAPD
Q.921	T1.602	LAPD Formats and Procedures

Frame Relay operates at the Physical and Data-link layers of the OSI Reference Model. It is also known as a "fast packet" technology because it keeps on sending the data without bothering about the error correction. If any error is detected in the transmitted frame, the frame is dropped. Error detection and correction is done at the endpoints of the WAN, which are also responsible for retransmitting the data. This makes the overall process of data transmission faster. Frame Relay requires a dedicated connection for data transmission. Figure 5.10 shows a frame relay network.

In WANs, Frame Relay technology is implemented using packet-switching technology with the help of virtual circuits, also known as Frame Relay *virtual circuits*. The virtual circuit creates a logical connection between two DTE terminals in a network for data transmission. It uniquely sets the path of communication

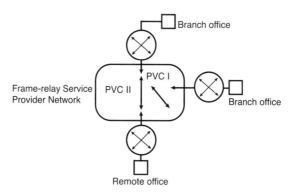

FIGURE 5.10   Transmission in a frame relay network.

between two endpoints in a network. It assembles the data to be transmitted into variable sized units called "frames," and then sends it across the *Frame Relay Packet-Switched Network* (PSN).

The Frame Relay virtual circuits are further categorized into the following two subcategories:

- Switched Virtual Circuits (SVCs)
- Permanent Virtual Circuits (PVCs)

Switched Virtual Circuits (SVCs)

SVCs refer to virtual circuits that are established and maintained during the time of data transfer. They are temporary in nature because the circuit is disconnected when the transmission is complete. An SVC lifecycle consists of the following four phases:

Phase 1—Call setup: In this phase, connection is established between two DTEs involved in data transmission.

Phase 2—Data transfer: In this phase, when the call setup establishes the connection, the data is transferred between two devices in a Frame Relay network.

Phase 3—Idle: In this phase, there is no data transfer between the DTEs. If there is no communication for a specified period of time, the connection is terminated between the DTEs. This phase initiates the call-termination phase.

Phase 4—Call termination: In this final phase, the virtual circuit or the connection is terminated between the communicating devices.

Permanent Virtual Circuits (PVC)

Unlike SVCs, PVCs are permanent in nature. This category of the Frame Relay virtual circuit is useful in situations where data needs to be transmitted frequently and consistently. Using PVCs, network users can specify the logical circuit and the bandwidth needed for the transfer of the data. PVC is responsible for managing the network traffic as well as the logical connections/circuits. PVC performs these tasks with the help of its constituents, presented as follows:

Endpoints: Nodes between which the data is transferred.

Committed Information Rate (CIR): Refers to the maximum number of bits transmitted in a specific time frame without incurring errors. The CIR of PVCs is decided by the physical capacity of the Frame Relay circuit.

Committed Burst Size (Bc): Specifies the maximum number of consecutive bits above the CIR that the PVC can carry without discarding the data.

Since the data transfer in PVC is continuous, it does not have a call setup or call termination phase. The other two phases are as follows:

Data Transfer: In this phase, data is transferred between two network terminals across the established virtual circuit.

Idle: In this phase, no data is transferred between the terminals, but the circuit is not terminated as in the case of SVCs.

After discussing the methodology of the data transmission, let us see in detail the form in which data is transferred in a Frame Relay PSN. In order to understand the functionality of a Frame Relay network, let us examine the structure of the frame format. Frame Relay uses the following two basic frame formats:

- Basic Frame Format version
- LMI version of the Frame Relay frame

Basic Frame Format Version

Figure 5.11 shows the frame format of a Basic Frame format.

As seen from Figure 5.11, a Frame Relay frame contains the followings fields:

Flag: Helps in streamlining the beginning and ending of the frame. The value of this field is represented either as a hexadecimal number (7E) or a binary number (01111110).

Address: This field maintains the following:

Data-Link Connection Identifiers (DLCI): Specifies the connection between the DTE terminal or device and a WAN switch. Its size is 10 bits. It

8 Bytes	16 Bytes	Variable	16 Bytes	8 Bytes
Flags	Address	Data	FCS	Flags

FIGURE 5.11   Frame format showing the basic frame format version.

helps the Frame Relay standard to uniquely identify a virtual circuit in a network. As DLCIs are unique to their particular Frame Relay link, the devices in the Frame Relay network can use different DLCI values to map to the same virtual connection.

Extended address: Used to indicate whether the byte in which the EA value is 1 is the last addressing field. If the value is 1, then the byte is indicated as the last in the DLCI octet.

C/R: This bit is not currently in use, but it is the next bit immediately after the most significant DLCI byte in the address field.

Congestion control: Contains the three bytes that help in implementing the congestion control mechanism in the Frame Relay network. This is implemented with these subcategories:

Forward-Explicit Congestion Notification (FECN): A single-bit field used by devices such as switches to inform DTEs devices about the congestion. An FECN value of 1 indicates the presence of congestion in the network.

Backward-Explicit Congestion Notification (BECN): A 1-bit field that functions as an FECN but indicates congestion in the direction opposite to the frame transfer from the source to destination.

Data: Contains the actual data to be transferred in the Frame Relay network. The maximum data length is 16,000 octets.

Frame Check Sequence (FCS): Ensures the integrity of the transmitted data. This value is first calculated at the source end and then recalculated at the recipient to check whether the frame received is the same as the one transmitted.

LMI Version of the Frame Relay Frame

LMI is a signaling protocol between the DTE devices and the Frame Relay switch. Figure 5.12 shows the frame format of an LMI Frame Format version.

The LMI frame format contains these fields:

Flag: Performs same functions as in the basic frame format.

LMI DLCI: Helps in identifying the frame as an LMI frame instead of the basic Frame Relay frame.

1 Byte	2 Bytes	1 Byte	1 Byte	1 Byte	1 Byte	Variable	2 Bytes	1 Byte
Flag	LMI DLCI	Unnumbered Information Indicator	Protocol Discriminator	Call Reference	Message Type	Information Elements	FCS	Flags

FIGURE 5.12 An LMI frame format showing different fields.

Unnumbered information indicator: Sets the final bit to the value 0.

Protocol discriminator: Contains a value indicating an LMI frame.

Call reference: Has a fixed value of 0. Currently, it does not serve any purpose.

Message type: Helps in labeling the frame into the following two message types:

> **Status inquiry message:** Allows the user devices to inquire about the status of the network.

> **Status message:** Corresponds to the status-inquiry messages.

Information Elements (IE): Contains a variable number of individual information elements (IEs). It has these subcategories:

IE identifier: Uniquely identifies the IE.

IE length: Indicates the length.

Data: Contains the encapsulated data (one or more bytes depending on the size of the data).

Frame check sequence: Performs the same functions as that of the basic frame format.

A recent advancement in the Frame Relay technology is the introduction of Voice Frame Relay Access Devices (VoFRADS). In its early stages Frame Relay was more famous for its data transmission features but with the advances in this field, Frame Relay now supports the transmission of voice data also.

The main advantages of a Frame Relay network are:

- Cost effectiveness
- Capability of handling heavy traffic load between the user and the end-to-end points
- A wide range of bandwidth availability
- Error control
- Effective congestion management features

X.25

X.25 is a telecommunication standard that sets the manner in which connections for data transmission are created and maintained in a packet-switched network. X.25 has a unique feature that allows workstations on diversified networks such as Banyan Vines and TCP/IP to communicate through a middle computer at the network layer. It operates effectively, regardless of the type of systems connected in the Internetwork. X.25 is used in lieu of dial-up or leased line mode of telecommunication.

X.25 is considered to be a Layer 3 methodology and is mapped to the lower three layers of the OSI reference model. These are the Physical, Data-link, and Network layers. Figure 5.13 shows the X.25 model mapping with the OSI layer.

X.25 uses certain networking protocols that are specific to its networks only. They are helpful in implementing and setting up X.25 PSNs. The protocols supported by X.25 are:

■ Packet Layer Protocol (PLP)
■ Link Access Procedure, Balanced (LAPB)
■ X.21bis

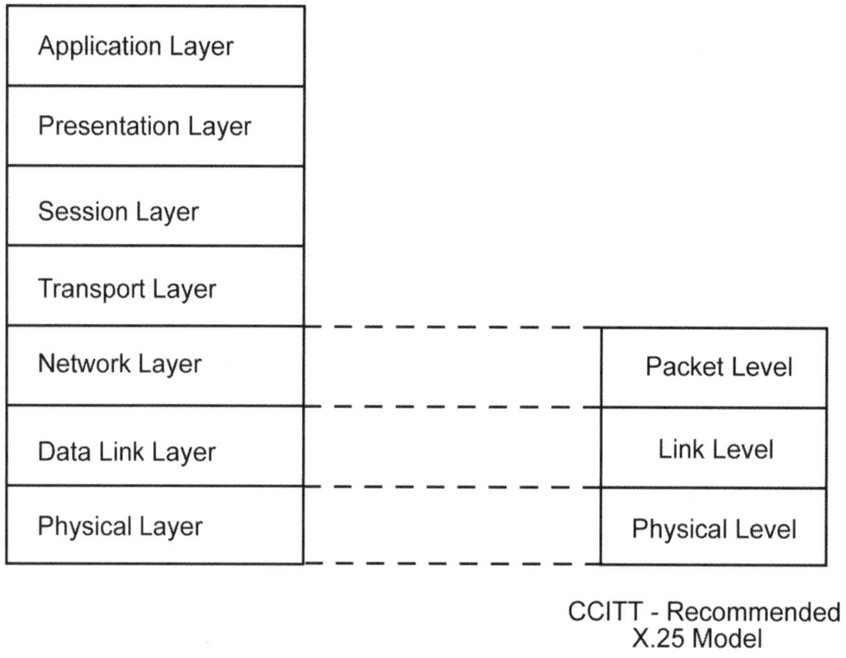

CCITT - Recommended
X.25 Model

FIGURE 5.13 X.25 model mapping with the OSI layer.

Packet Layer Protocol (PLP)

PLP is directly linked with managing packet exchanges across virtual circuits between two DTEs. For managing, it uses five modes, as shown in Table 5.5.

TABLE 5.5 PLP Modes

Mode	Description
Call Setup mode	Used to form SVCs between two interconnecting terminals. The circuit is established before sending the data.
Data Transfer mode	Used while transferring data between two DTEs. In this mode, PLP performs segmentation, reassembly of packets, padding, and error and flow control. These functions are common to both SVCs and PVCs.
Idle	No data transfer takes place here. This mode is operational in SVCs only.
Call-clearing mode	Used to clear all the circuits or connections between two communicating ends. This mode is used only in SVCs.

PLP also sets certain fields for managing the X.25 network data packets. These are as follows:

General Format Identifier (GFI): Checks the general format of the data. It performs various checks on the data whether the data packet is carrying the data or the flow control information.

Logical Channel Identifier (LCI): Streamlines the virtual circuit across the DTE and DCE interface.

Packet type identifier: Identifies the data packet as one of the 17 different PLP packet types.

User data: Consists of the encapsulated information pertaining to data of the upper layers of the OSI Model.

Link Access Procedure, Balanced (LAPB): LAPB is associated with framing of the data packet and error checking. This LAPB frame contains these fields:

Flag: Indicates the startpoint and endpoint of the frame.

Address: Determines whether the frame is carrying a command or a response.

Control: Deals with signaling and control information related to the frame.

Data: Consists of the upper-layer data contained in a PLP packet.

FCS: Responsible for error checking and verifying whether the frame has reached its destination intact.

Figure 5.14 depicts an LAPB frame structure.

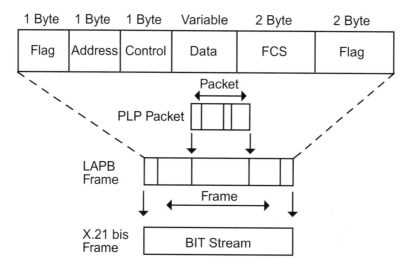

FIGURE 5.14 An LAPB frame structure showing different fields.

X.21Bis

This is a Physical layer protocol used in X.25. It is related to the electrical mechanism and procedures of the physical medium. This protocol also handles the activation and deactivation of the physical medium that is used to connect the DTE and DCE equipment.

Figure 5.15 illustrates mapping of X.25 protocols and OSI layers.

Apart from the usual network devices such as routers and switches categorized into DTEs and DCEs, the X.25 technology uses certain devices that are specific to the X.25 standard. These are as follows:

Packet-Switching Exchange (PSE) devices: These devices are switches specific to the X.25 protocol and are used to compose the bulk of the carrier's network. They are responsible for transmitting data packets called "frames," between DTEs in an X.25 PSN. Figure 5.16 shows an X.25 setup.

Packet Assembler/Dissembler (PAD): These devices are specific to the X.25 PSN and are used with DTEs, which are Character User Interface (CUI) based. The PAD is used between DTEs and DCEs to perform the following functions:

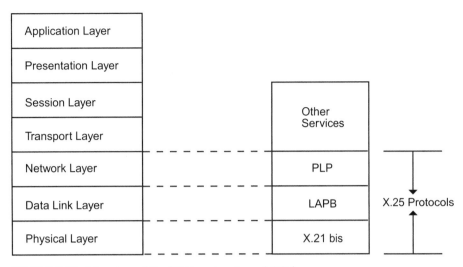

FIGURE 5.15 Mapping of the X.25 protocols and OSI layers.

Buffering: PAD stores the data from the DTE terminal until the device is ready to process the data.

Assembling: After buffering, when the device is ready, PAD assembles the data and forwards it to the DCE device. It also adds the X.25 header to the data.

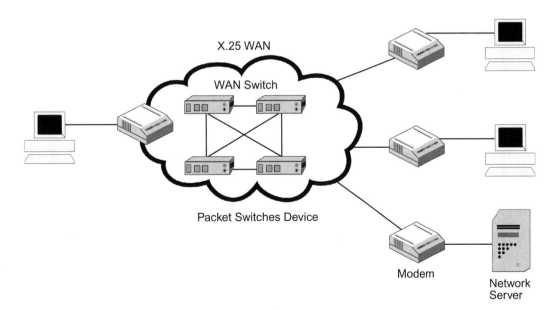

FIGURE 5.16 An X.25 setup showing DTEs and an X.25 PSN.

Disassembling: On receiving the data, PAD disassembles it before re-forwarding it to the DTE. At the time of disassembling, it also removes the X.25 header from the data.

Benefits of X.25

X.25 offers the following benefits:

Reliability: X.25 offers reliable transfer of data across the network.

Quality: X.25 offers quality data transmission on both analog as well as digital media.

Fault diagnosis: X.25 employs error checking and detection to help diagnose the fault in the data flow.

Switched Multimegabit Data Services (SMDS)

SMDS as a telecommunication service was first set up by the IEEE 802.6 and was used by the Bell Communications Research (Bellcore). SMDS can transmit data across any kind of network platform, enhancing the flexibility features of its services. Figure 5.17 shows an SMDS WAN setup.

SMDS consists of these components:

Customer Premises Equipment (CPE): This is a device located at the customer site and maintained by them. CPE includes equipment such as per-

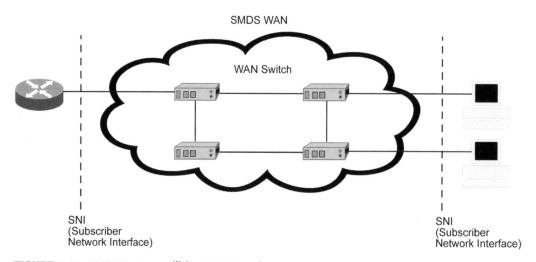

FIGURE 5.17   A WAN setup utilizing SMDS service.

sonal home PCs and connective devices such as routers, multiplexers, and modems.

Carrier: Refers to a physical or hardware medium that connects the various endpoints across a WAN network. SMDS includes high-speed WAN switches.

Subscriber Network Interface (SNI): Refers to an interface between the CPE and the carrier. It helps to connect both the ends together for the successful flow of data packets.

Let us learn how information flows in an SMDS implemented network.

For data transmission, SMDS uses the SMDS Interface Protocol (SIP) to ensure information flow.

SMDS Interface Protocol (SIP)

This is a protocol to establish communication between CPEs and SMDS-specific carrier equipment. SIP provides connectionless service that helps SNI to establish connection between CPEs and carriers. SIP has its foundations in *Distributed Queue Dual Bus* (DQDB), an IEEE 802.6 standard for cell relay transmission in MANs.

SIP is further subdivided into three layers according to the functions performed by each layer. These layers are as follows:

- SIP Level 1
- SIP Level 2
- SIP Level 3

SIP Level 1

SIP Level 1 operates at the Physical layer of the OSI reference model. It helps in implementing and connecting to the physical medium; the medium can be DS-1 (T1/E1) or DS-3 (T3/E3). This layer also consists of Physical Layer Convergence Protocol (PLCP) sublayers. PLCP helps to decide how SIP Level 2 cells are to be arranged with reference to the physical media (DS-1 or DS-3).

SIP Level 1 cells have formats containing the following fields:

Access control: Contains different values depending upon the user information received from the CPE. It is 8 bits in length.

Network control information: Contains a specific value that notifies whether the Protocol Data Unit (PDU) has any user information or not. It is 32 bits in length.

Segment type: Differentiates whether the received cell is the first, middle, or the last in the PDU. It is 2 bits in length. The four set segment values are:

00: Stands for message continuation

01: Stands for message end

10: Stands for message start

11: Stands for message is single segmented

Message ID: Associates Level 2 cells with the Level 3 PDU. It is 14 bits in length.

Segmentation unit: Contains the data of the cell. It is 352 bits in length.

Payload length: Indicates the total number of bytes stored in the segmentation unit. It is 6 bits in length.

Cyclic Redundancy Check (CRC): Related to error checking and detection. It is 10 bits in length.

SIP Level 2

SIP Level 2 operates at the Data-link layer (Layer 2) of the OSI Reference Model. Level 2 acts on the PDUs forwarded by SIP level 3. It segments the PDUs into fixed-sized cells (the size of the cell is 53 octets). PDUs contain the source and destination addresses.

SMDS addresses are 10-digit values relating to addresses.

NOTE

When the formatting is complete, they are sent to the SIP Level 1 for placing them on the physical medium for transmission.

SIP Level 3

SIP Level 3 operates at Layer 3 of the OSI reference model. It interacts directly with the data packets called SMDS Service Data Units (SDUs). The SDUs are put into Level 3 header and trailer forming a frame, which is referred to as Level 3 Protocol Data Unit (PDU). After PDUs are formed, they are passed to the SIP Layer 2.

In an SMDS network, the CPEs pass the information to the carrier. The SMDS protocol converts the data into the required format (say cells). When the data has undergone formatting and has been converted into cells, the carrier passes it to the endpoint across the network with the help of the destination address specified in the cell.

Benefits of SMDS

Some of the benefits of SMDS technology are:

High-speed data transmission: Offers high-speed data transmission due to its connectionless feature

Network flexibility: Offers flexibility and can be implemented on any network platform for data transmission

Network scalability and expandability: Offers the option of accommodating more users without disturbing the flow of data transmission

Asynchronous Transfer Mode (ATM)

ATM is a data transmission technology specifically used for voice and video transmission. It is based on cell-switching methodology. ATM uses virtual connections called SVCs and PVCs as in Frame Relay networks. ATM incorporates some of the features of packet-switching methodology. It covers the digital data into fixed-sized byte "cell" units. Each cell consists of 53 octets. These 53 octets are further subdivided into cell header information (first 5 bytes) and user information (48 bytes). When the data is organized in the cells, it is transmitted over a physical medium that uses digital signaling technology. The data transfer rate is very fast, ranging from 155 Mbps to 10 Gbps because ATM uses a digital medium. Figure 5.18 depicts an ATM setup.

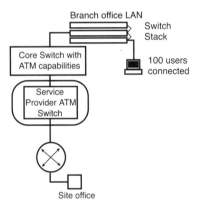

FIGURE 5.18 Data transfer in an ATM setup.

ATM Devices

There are certain devices designed specifically for ATM networks, which serve as a strong backbone in data transmission. These ATM-specific devices are as follows:

ATM endpoints: Refers to the end system, which contains an ATM network interface adaptor. Examples of ATM endpoints are workstations, LAN switches, routers, and DSUs.

ATM switches: Refer to switches specific to ATM WAN networks. Their functions are similar to WAN switches. Figure 5.19 illustrates various devices used in an ATM network.

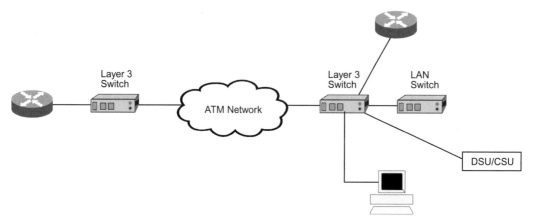

FIGURE 5.19   Various devices used in an ATM network.

ATM network interface: Used to interconnect two ATM devices across point-to-point links. They are categorized as follows:

User Network Interface (UNI): Acts as an interface between the ATM endpoints and ATM switches. The UNI is divided into private and public UNIs. The Public UNI connects the ATM endpoints to public switch (a public switch spans over two or more LANs). In case of private networks, UNI connects an ATM endpoint and a private switch.

Network Node Interface (NNI): Acts as an interface between two ATM switches.

ATM Operations

ATM technology uses certain processes, functions, and protocols for the successful transmission of data. As mentioned earlier, data in ATM networks is transmitted in the form of fixed-sized cells. An ATM cell is 53 octets in size and structurally has a cell header and user information technically called *payload*.

The cell header format can use either a UNI or NNI format.

A cell header in the UNI format has the following fields:

Generic Flow Control (GFC): Responsible for flow control of the data from the endpoint devices to the ATM switch

Virtual Path Identifier (VPI) and Virtual Channel Identifier (VCI): Form the virtual connection across the ATM network

Payload type: Indicates whether a cell contains the user information or control information

Cell loss priority: Indicates whether the cell has to be discarded in case of network congestion

Header error control: Contains CRC information

Figure 5.20 illustrates a UNI cell header format.

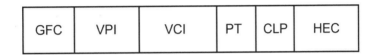

FIGURE 5.20   A UNI format showing cell headers.

The NNI cell header format has the same fields as that of a UNI cell header format. The only difference is that the Generic Flow Control (GFC) of a UNI format is replaced by an expanded VPI space in an NNI format. Figure 5.21 illustrates an NNI cell header format.

FIGURE 5.21   An NNI cell header format showing different fields.

ATM contains a reference model of its own to set the various functions and protocols used to ensure successful data transmission over ATM networks. The ATM reference model corresponds to the Physical and Data-link layers of the OSI reference model. An ATM Reference model consists of these layers:

- ATM Physical layers
- ATM layer
- ATM Adaptation Layer (AAL)

ATM Physical Layer

This layer is responsible for managing medium-dependent transmission. The physical layer is further subdivided into:

Physical medium dependent sublayer: Specifies the physical medium format for data transmission. It synchronizes transfer and reception of data packets by maintaining continuous flow bits with their timing information.

Transmission Convergence (TC) sublayer: Responsible for maintaining the flow of cells by checking for header error control in case of trouble in sending the data, and maintaining the cell flow by preventing the cells from going to idle. This sublayer divides the data cells into frames for further transmission.

ATM Layer

This layer starts a connection (cell circuit) and forwards the cells across the ATM network. It helps in multiplexing and demultiplexing the cells and also adds the header before as well as after the cell is forwarded to AAL.

ATM Adaptation Layer (AAL)

This layer creates isolation between the high-layer protocols from the ATM processes and procedures by converting the information from high layers into cells. The two sublayers of AAL are as follows:

Convergence sublayer: Divides the frame into 53-byte cells and forwards it to the destination for reassembly.

Segmentation and reassembly sublayer: Segments the frames into ATM cells at the sending end or node and reassembles it into proper format on the recipient end.

Figure 5.22 depicts various layers in an ATM Reference Model.

Benefits of ATMs

The benefits offered by ATMs are as follows:

High performance: ATM networks transfer data at high speed.

Support for multimedia applications: Applications using voice and video data can be transmitted easily using ATM cells.

Interoperability: A common LAN/WAN architecture makes it possible to be implemented from one workstation to another irrespective of the software and hardware used.

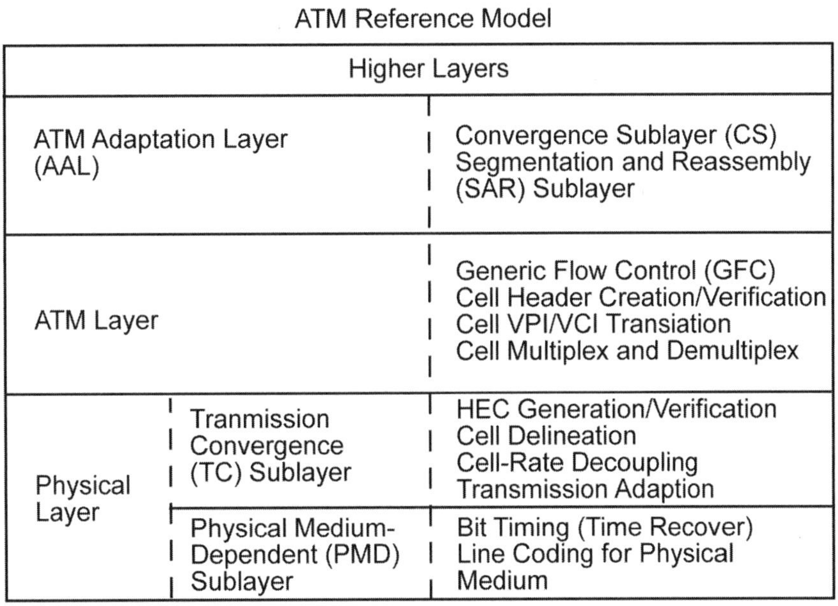

ATM Reference Model

Higher Layers		
ATM Adaptation Layer (AAL)		Convergence Sublayer (CS) Segmentation and Reassembly (SAR) Sublayer
ATM Layer		Generic Flow Control (GFC) Cell Header Creation/Verification Cell VPI/VCI Transiation Cell Multiplex and Demultiplex
Physical Layer	Tranmission Convergence (TC) Sublayer	HEC Generation/Verification Cell Delineation Cell-Rate Decoupling Transmission Adaption
	Physical Medium-Dependent (PMD) Sublayer	Bit Timing (Time Recover) Line Coding for Physical Medium

FIGURE 5.22   ATM reference model showing different layers.

Tables 5.6 and 5.7 list important comparisons among various technologies.

TABLE 5.6   Technology Classifications

Category	Analog Dialup	ISDN	Leased line
Maximum Bit Rate	33 to 56 Kbps	64 Kbps to 2.048 Mbps	<2.5 Gbps
Charge	Time, distance	Distance, capacity	Distance, capacity
Connection Type	Dial-up	Dial-up	Dedicated
Speeds	Slow	Moderately fast	Very fast
Bandwidth	Limited/shared	Shared	Full
Control	Low	Low	High
General Characteristics and Use	SOHO, backup	Higher bandwidth than dialup, DDR	High speed, always on High cost to enterprise

TABLE 5.7   Technology Classifications

Category	X.25	Frame Relay	ATM
Maximum Bit	<48 Kbps	<4 Gbps	<156 Mbps
Rate Charge	Volume	Capacity	Capacity
Connection Type	Switched fixed	Virtual switched or permanent	Cell switched, SVC or PVC
Speed	Moderately slow	Very fast	Fast
Bandwidth	Shared	Shared	Shared
Control	Low	Medium	Medium
General Characteristics and Use	High reliability	High speed, Bandwidth on demand point-to-multipoint. Medium cost to enterprise	Voice, video, and data integration

WAN PROTOCOLS

In this section, we will discuss various WAN protocols that are used in data transmission. These are:

- Synchronous Data Link Control (SDLC)
- High-Level Data Link Control (HDLC)
- Point-to-Point Protocol (PPP)

Synchronous Data Link Control (SDLC)

The *SDLC* protocol is used for transmission of data based on the synchronous mode of bit-oriented operation. SDLC can be used in point-to-point as well as multipoint links with both half-duplex as well as full-duplex modes of transmission over bounded or unbounded media. SDLC was developed by IBM in 1970 to use in the System Network environment scenarios. Later, the International Standards Organization (ISO) modified the SDLC standards to be used in the International markets.

The SDLC protocol standard is based on the primary-secondary station model of communication. In this model, the primary station is the mainframe or the main server. Other devices such as workstations and network devices such as routers and bridges are the secondary devices (each secondary device has a unique address). These two are connected either through a point-to-point connection or a multipoint connection. SDLC is more successful in private networks with dedicated lines as media.

The primary workstations and secondary workstations can be linked or connected with each other through the following four basic methods:

Point-to-point link: Specifies two workstations—primary and secondary.

Multipoint link: Specifies a primary station and multiple other stations.

Loop link: This link specifies a ring-type structure, in which the primary link is connected with the first secondary link and last secondary link. The other secondary links are connected with each other and communicate to the primary link via messaging.

Hub-go-ahead: This link has two major channels—inbound and outbound. Primary stations use outbound channels for connecting or communicating with the secondary stations in the network. Secondary devices use the inbound channel for communicating.

In SDLC, the data information flows in the form of frames. Figure 5.23 shows an SDLC frame format.

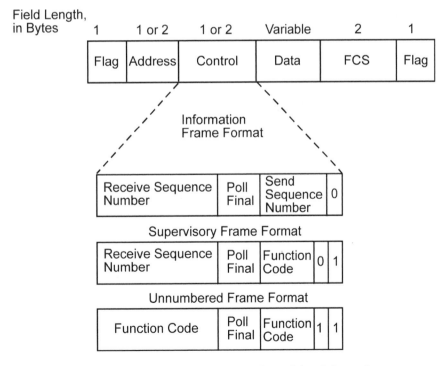

FIGURE 5.23 An SDLC frame format showing flow of data information.

Different fields in the frame format are as follows:

Flag: Helps to indicate the starting and ending of the frame and error checking related to the frame.

Address: Consists of the destination address of the secondary terminal or device. The specified address can be a single secondary address, a group address, or a broadcast address.

Control: Contains the following subdivisions:

> **Information Frame:** Consists of the upper layer and control information

> **Supervisory Frame:** Handles the status of the frame (whether it has reached its destination safely). It can also suspend or request frame transmission.

> **Unnumbered Frame:** Used to start the initialization phase for secondary devices for further data transmission.

Data: Contains the user information.

FCS: Checks if the frame at the recipient end is the same as that sent by the source.

SDLC served as a precursor for other network protocols, such as HDLC and LAPB.

High-level Data Link Control (HDLC)

HDLC is a protocol used in transmitting data information between endpoints across networks. The data that is to be sent is arranged in units called frames.

HDLC is based on the SDLC protocol, a networking protocol that was developed by IBM for transmission of data. It is similar to SDLC and can be referred to as a subset of SDLC. HDLC follows the primary terminal-secondary terminal model. As per the model, the primary terminal is the mainframe or the server and the secondary terminals are the user-end terminals such as workstations and intermediate WAN devices such as bridges and WAN switches. The primary terminal is connected to the secondary terminals in a point-to-point link or a multipoint link. The frame structure in HDLC is very similar to the frame format structure set in SDLC. Figure 5.24 shows an HDLC frame format.

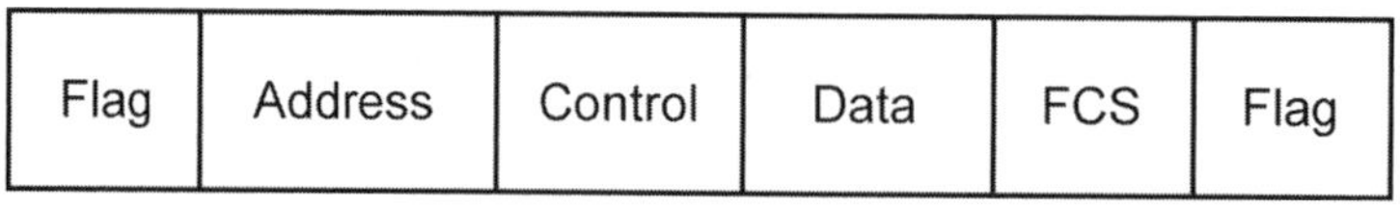

Flag	Address	Control	Data	FCS	Flag

FIGURE 5.24   An HDLC frame format showing different fields.

The HDLC frame has the following fields:

- Flag
- Address
- Control
- Data
- FCS

The functionality of all these fields is the same as that of the SDLC frame fields. The starting and ending points in the frame are marked by the Flag field. The address field contains the destination address of the recipient endpoint. These addresses can be a single address or the group address.

The main difference between the HDLC and the SDLC network protocols is the way in which secondary equipment interacts with the primary equipment for transferring data. SDLC recognizes only one kind of transfer mode, that is, the *Normal Transfer Mode* (NTM) while HDLC supports the following three transfer modes:

Normal Response Mode (NRM): In this mode, the communication between secondary devices and the primary devices can be initiated only when the primary devices grant permission.

Asynchronous Response Mode (ARM): In this mode, the communication can be initiated between secondary devices and primary devices without any prior permission.

Asynchronous Balanced Mode (ABM): In this mode, any device, either primary or secondary, can initiate communication without any prior permission. That means even secondary devices initiate communication to the primary device.

Point-to-point Protocol (PPP)

PPP is a protocol that helps in connecting a terminal to the other terminal in the network over serial point-to-point links. It is located in the Data-link Layer of the OSI reference model. PPP operates in the full-duplex mode over any kind of physical media whether it is copper or optical fiber. PPP performs functions such as the following:

- Assignment and management of IP addresses of the equipment in the network
- Encapsulation of bits
- Multiplexing of network protocols
- Link configuration between endpoints in the network
- Error checking

The two major subdivisions of PPP are:

Link Control Protocol (LCP): Helps in establishing, configuring, and testing of various links in the network

Network Control Protocol (NCP): Helps in configuring different network layer protocols used for internetwork connections

Let us see how PPP operates in achieving the functions mentioned here. PPP first sends an LCP frame to initiate, establish, and test the terminal link connection. After the initialization process is over, the next step is to send the NCP frames to configure the Layer 3 protocols. When the groundwork of establishing a pathway between terminals is complete, data can be sent over the established connection. This path will remain active until the NCP link closes the frame by sending the appropriate frame field indicated at the end.

Figure 5.25 shows a PPP frame format:

Flag	Address	Control	Protocol	Data	FCS

FIGURE 5.25 A PPP frame format showing different fields.

Various fields in the frame format are as follows:

Flag: Used to indicate the start or end of the frame. The size of this field is 1 byte.

Address: Contains the standard broadcast address because it does not support assigning individual addresses. The size of the field is 1 byte.

Control: Gives a call for transmission of the user data. The size of the field is 1 byte.

Protocol: Identifies the specified related protocol in the information field of the frame.

Data: Contains the datagram for the protocol mentioned in the protocol field. The maximum size can range up to 1500 bytes.

Frame Check Sequence (FCS): Helps in error detection. This field is 16 bits in size.

Some features unique to PPP are:

Address Notification: In a network using a PPP link, a server can pass its IP address to the client linked to it.

Use of multiple protocols: PPP supports use of multiple protocols on the same link.

Link Monitoring: PPP checks the functioning of other links in a network.

SUMMARY

In this chapter, we learned about the hardware necessary to build a cost-effective, high-bandwidth WAN. We also reminded you about provisioning media based on the area covered, clients' budget issues, traffic patterns, bandwidth requirements, and scalability. In the next chapter, we move on to LAN segmentation using different devices.

POINTS TO REMEMBER

- A WAN network should be designed to provide maximum bandwidth, minimum tariff cost, and maximum service to the users.
- Important design considerations while designing WANs are backbone load determination, bandwidth distribution, and distance limitation calculations.
- Different types of WANs include WAN routers, switches, access servers, modems, terminal adapters, CSU/DSU, Data Terminal Equipment (DTE), and Data Communications Equipment (DCE).
- A WAN Access Server performs authentication, authorization, and accounting.
- CSU/DSU converts the digital frame used in LANs into a frame format appropriate for a WAN-specific environment.
- WAN media options include leased lines, fiber links, and wireless.
- Leased lines provide an ideal solution for bandwidth-demanding low-latency multimedia applications.
- The international equivalent of SONET is Synchronous Digital Hierarchy (SDH).
- Different types of DSL are ADSL, CDSL, G.lite or DSL lite, and HDSL.
- ADSL enables upstream and downstream data transfer.
- HDSL follows a symmetric mode of transmission wherein an equal amount of data is transferred in both directions.
- The components of a leased line include CPE, Carrier-end Modems, and Transmission Medium.
- In case of fiber links, data is transmitted through optical fiber cables instead of conventional copper cable.
- Circuit-switching services are also called Time Division Multiplexing (TDM) services, which involve data transfer in fixed time slots.
- Different ISDN-specific devices include terminals, terminal adapters, and network terminal devices.
- The three types of services offered by ISDN are Basic Rate Interface (BRI), Primary Rate Interface (PRI), and Broadband-ISDN (B-ISDN).
- Frame Relay is a technique used for data transmission between LANs and endpoints of WANs.

- An X.25 standard is used in lieu of dial-up or leased line mode of telecommunication.
- X.25 supports Packet Layer Protocol (PLP), Link Access Procedure, Balanced (LAPB), and X.21bis.
- ATM covers the digital data into fixed-sized cell units formed by 53 octets.
- A cell header format can use either a UNI or NNI format.
- The fields of an HDLC frame include flag, address, control, data, and FCS.

6 | LAN Segmentation

With more and more bandwidth-demanding and network-intensive applications such as multimedia applications, and increasing numbers of users connected in a LAN, upgrading the backbone network would not be the best solution. To solve this problem of insufficient bandwidth, there are at least two options. One is to install faster network technologies such as Asynchronous Transfer Mode (ATM) or Fiber Distributed Data Interface (FDDI). This would mean incurring more expenses to manage and maintain the network. Another feasible approach would be to segment the entire network into smaller groups using bridges, routers, and switches. Implementing these technologies would lessen network bottlenecks and frame losses. This process of breaking down a large network into smaller segments is called *LAN segmentation*.

LAN SEGMENTATION

LAN segmentation is the process of dividing a large LAN into smaller groups or subnetworks called *segments*, for the purpose of effective network management. Network or LAN segmentation is needed when a network cannot be expanded further without affecting the network performance. LAN segmentation solves many network-related problems such as congestion and bandwidth availability. It helps in extending the network and improving security, and in separating network problems by dividing the network into smaller manageable broadcast domains.

Some of the advantages of LAN segmentation are as follows:

Reducing traffic congestion: Due to lower numbers of users per segment, the data flow between devices will be faster, so fewer collisions will occur.

Increasing bandwidth: Dividing a LAN into smaller groups means that the number of users who demand bandwidth on the LAN segment is lower. Therefore, bandwidth distributed per user will be greater. This should allow the handling of even the most bandwidth-hungry applications.

Enhancing security: Segmentation can stop the data packets from spilling over a particular segment; that is, the data will be circulated within a given segment without involving other segments. Local traffic will remain on the local segment and increase privacy and secrecy of the data within a segment.

Isolating network problems: Segmentation can also help stop spilling over of problems (hardware as well as software failures) from one segment to another. This means that failure at one point would not stop the functioning of the entire network; the other segments will continue working. Therefore, troubleshooting the problem pertaining to a particular segment is easier.

Implementation of LAN is a critical process. It can be implemented through various hardware devices and technologies. A LAN can be segmented in the following two ways:

- Physical segmentation
- Network switching

PHYSICAL SEGMENTATION

Physical segmentation of LANs can be implemented using hardware devices such as bridges, routers, and switches. These devices are used to create smaller subnetworks or broadcast domains out of the bigger networks.

LAN Segmentation Using Bridges

Bridges are used to interconnect two or more LAN segments. They are responsible for forwarding data frames based on MAC addresses from one segment to another. They act as store-and-forward devices. Bridges operate at Layer 2 of the OSI model and can create two or more LAN segments. Bridges are responsible for extending LANs to overcome physical limitations and support LAN segmentation to help remove traffic congestion. Since bridges do not process networking protocols, they support all the protocols, offering very good throughput.

Bridges are intelligent devices that apply filtering on frames and can be programmed such that they can reject frames from other network segments to help control traffic. The frame contains the MAC address of the destination port so the bridge can directly forward the frame to that particular node. There are two types of bridges, local and remote. *Local bridges* provide connection between multiple LAN segments in the same area. *Remote bridges* provide connections between multiple LAN segments placed at different locations.

The two main bridging techniques used in internetworking environments are as follows:

Transparent bridging: The bridges decide the frame path by reading the MAC address and using the bridging table. After reading the address, the bridge sends the frame to the destination port. In case no address is found in the table, the frame is broadcasted to all ports. This type of bridging is used in Ethernet.

Source-route bridging: An example of source-route bridging is a Token Ring LAN. In this case, the sending node or the port decides the route. Bridges decide on the path of the frame so that the frame reaches its destination quickly.

As mentioned, bridges forward frames on the basis of the MAC address contained inside the frame. In cases in which the MAC address is unavailable, the bridge resends the frame, creating a loop (that is, sending the frame again and again). Looping is harmful for a network because it consumes a lot of bandwidth and slows its performance. To overcome the problem of looping data packets, bridges use the Spanning Tree Protocol (STP) to eliminate the loops.

STP is a link management protocol that provides path redundancy for the data transmission and eliminates looping in the network. STP prevents looping by providing only one active path at a given point in time between network devices. In addition, STP uses the Spanning Tree Algorithm (STA) for implementation. The STA provides an efficient path for the bridge to forward the frame to avoid looping. If collision occurs, STA again calculates the best possible path for the bridge to forward data. STA encloses the information of the best possible route in the *Bridge Protocol Data Unit* (BPDU). BPDUs are packets containing information about the port addresses and priorities assigned to the frames. BPDU messages are exchanged across the bridge devices to detect and avoid loops in the network. Some benefits of using bridges for LAN segmentation are as follows:

■ Provides extendibility to LAN segments
■ Connects multiple LAN segments (locally as well as remotely)

■ Handles multiple protocols

■ Eliminates looping by using STP in order to reduce traffic congestion

LAN Segmentation Using Routers

Routers interconnect various LAN segments so that communication and data transfer can take place between different LANs. A router functions at the Network layer of the OSI model. Routers provide multiple paths for communication between different LAN segments with the help of routing protocols and an internal routing table. It also uses the IP address of the destination system to decide the route of the frame. If the frame is to be sent to the node on the same LAN segment, the router will send it directly; otherwise it will use its routing table. Some of the benefits of using routers are as follows:

Media transitioning: Routers connect LAN segments using different media types. Let's say, for example, we have a network that encompasses both an Ethernet and a Token Ring LAN. Routers can be used to connect these LANs. Since routers are Layer 3 devices, they can take care of Layer 3 address translations and fragmentation requirements.

Packet filtering: Routers can filter packets between different segments.

Backbone for VLANs: Routers help in connecting and communicating among various VLANs.

Broadcast controlling: Routers prevent broadcasts from spilling over to other domains, that is, they help in keeping data of a particular broadcast domain within its own domain. Routers also provide various services for each LAN segment. These are the examples of services provided by the routers for a variety of protocols:

IPX: SAP table updates.

AppleTalk: ZIP table updates.

Network management: SNMP queries.

IP: Proxy ARP and ICMP.

LAN Segmentation Using Switches

Switches help in connecting multiple LAN segments to form a single robust network. Switches function at Layer 2 of the OSI model. They forward the frame to the destination port based on the MAC address contained in the frame.

Switches use any of the following three methods to forward data in a frame:

Cut-through: Streamlines data such that the first data packet exits the switch before the rest of the data in the frame enters the switch. In this method, no error checking is performed; therefore, corrupted frames may take up some of the bandwidth.

Store-and-forward: Copies the entire data frame into the memory of the switch before sending it to the destination address. In this method, a complete CRC is done before forwarding. If errors are detected, the frame will not be forwarded. This method prevents the bandwidth consumption that can occur with the cut-through method.

Fragment-free: This method is a combination of both the cut-through and store-and-forward methods. The switch will store only the first 64 bytes of the data frame in its memory, but it will perform a CRC check. Most of the errors occur in this part; therefore, bad frames are detected and dropped.

Some of the advantages of LAN switching are:

- Switches help connect multiple LAN segments with collision-free communications between the end systems.
- In a traditional LAN environment, bridges and hubs are used to segment the network and transmit the data. With bandwidth demand on the rise, LAN switching is a good solution. As compared to bridges, switches have a higher number of ports because they divide the network into several dedicated channels. This results in multiple independent data paths, increasing the throughput of the switches.
- Another advantage of switches is that they are self-configuring, which helps in minimizing network downtime. Deploying LAN switches is easy because it does not require changing existing hubs, NICs, or cabling. To sum up, switches provide higher port density at lower cost as compared to bridges.
- LAN switching has also solved the bandwidth problem through *microsegmentation*. Microsegmentation is a concept of dividing a network into private or dedicated segments, that is, allocating one workstation per segment. The major benefit of microsegmenting is that each node gets access to the full bandwidth instead of having to share it with other users on the network. As a result, chances of collision are much lower, especially in a full-duplex environment.
- Switches can maintain multiple connections at the same time.

NETWORK SWITCHING

It is also possible to divide a network logically. Logical segmentation can be achieved through switches at Layers 2 and 3 of the OSI model as in case of VLANs.

The major advantage is that this allows dynamic segmentation and formation of network groups without altering the physical layout of the network. Now let us discuss the network-switching concept with VLANs.

Virtual Local Area Network (VLAN)

VLANs or Virtual LANs are a group of LANs logically connected but physically located on different segments of the LAN.

A VLAN is a segmented network, a network divided on the basis of applications, project teams, and departments that is segmented on a logical basis rather than on a physical basis. VLANs are used in campus LAN environments. Figure 6.1 shows a VLAN network.

Figure 6.1 is a good example of a VLAN. It shows how workstations grouped under departments such as Engineering, Marketing, and Accounting, although physically separate, are connected logically with the help of routers and switches.

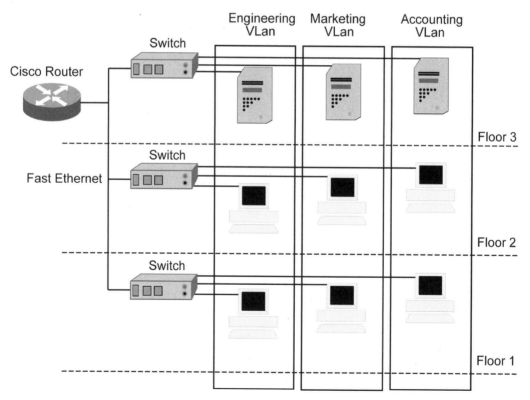

FIGURE 6.1 A campus VLAN environment showing different departments.

There are two types of VLANs, static and dynamic. Let us discuss each of these types.

Static VLANs

The static VLAN is the most common and widely used VLAN because of manageability and cost factors. Static VLANs are configured by network administrators by assigning physical ports to each node in a VLAN segment. For example, an administrator can join together 10 nodes in the marketing division (although located physically apart) into a VLAN segment by assigning port addresses to these nodes from 1 to 10. Static VLANs are easy to configure and monitor, but at the same time, the major drawback is that static VLANs do not offer user mobility. This means if a new user joins a different broadcast domain, the administrator has to reconfigure the entire VLAN.

Dynamic VLANs

Dynamic VLANs are not bound with physical ports as in the case of static VLANs. They are based on the properties of OSI Layers 2, 3, or 4. With Layer 2, nodes in a VLAN segment can be configured using the connecting MAC addresses. With Layer 3, it is done on the basis of IP addressing. In Layer 4, configuration is based on the application itself. Therefore, a VLAN switch will dynamically assign a node to a particular VLAN segment by reading the specific property (MAC address, IP addressing, or application) of the node or the device. Table 6.1 compares static VLANs with dynamic VLANs.

TABLE 6.1   Static and Dynamic VLANs

Static VLANs	Dynamic VLANs
Configured by the administrators	Can be configured dynamically
User mobility (additions and deletions) not easy	User mobility easy because any additions or deletions are autoconfigured
Bound by physical ports	Configured on the basis of MAC addresses and IP addressing

VLAN Operation

In VLANs, data is forwarded or carried in the form of frames. When a LAN bridge has to forward the data to the destination endpoint, it adds a tag to the frame specifying the destination VLAN port where the data is to be delivered (if we have more than one VLAN in a network). This process is called *frame tagging*.

Frame tagging is a method to keep track of the number of users/nodes belonging to a particular VLAN in a network as well as the frames traversing through the network. Frame tagging is useful in identifying the VLAN membership of a frame; that is, to which VLAN segment it belongs. Since there are many ports in a particular VLAN and many VLANs in a network, a record of ports and destination addresses is maintained in the *filtering database*. The bridge maintains the database so that it forwards the data frame to that particular frame of a VLAN for which it is meant.

A protocol is needed to handle all the configured VLANs in a network because in a network, there is more than one VLAN segment. This protocol is called the *VLAN Trunk Protocol* (VTP). VTP is responsible for managing additions and deletions of VLANs in a network and renames VLANs to avoid duplication. VTP is not required if there is only one VLAN in a network. Some functions of VTP are as follows:

- Maintaining consistent VLAN information in all switches in a domain
- Tracking and monitoring of various VLANs in the network
- Detecting and reporting new VLANs
- Configuring new VLANs

Some of the benefits of VLANs are as follows:

Simple manageability: Expanding a network is easy because any new additions can be done through the management console rather than the wiring closet.

Flexible network segmentation: Grouping the nodes and the resources in the same domain is easy. For example, computers and printers in a marketing division can be grouped together in a VLAN segment without concern to the physical location of nodes.

Increased performance: Grouping all the related nodes together helps the network administrators build broadcast domains with fewer users. With fewer users, they can have dedicated bandwidth assigned to them and increase the performance of a complete VLAN segment.

Enhanced network security: Securing the group of nodes in VLAN is easy because these groups are surrounded by *virtual boundaries*. These boundaries act as a safety wall and restrict access to each VLAN segment.

80/20 RULE—20/80 RULE

The 80/20 rule says that 80 percent of the traffic is to be contained in the LAN/LAN segment and 20 percent of traffic should leave the network through routers. The reason why such a rule came into being was that in a traditional network model, each department had its own central server. In the current scenario, with central servers placed anywhere on the World Wide Web, the traffic pattern is changing. Currently, 20 percent of the traffic remains inside the network and 80 percent leaves the network as the file exchange is sometimes within the extranet.

QUALITY OF SERVICE (QOS)

QoS is a set of tools and techniques to measure and maintain the quality of data transmission and network resources. This helps in avoiding the following problems:

Delay: An increase in the total time taken by the data frame to reach its destination endpoint or node across a network

Lower throughput: A decrease in the rate at which the data is transferred through the network

Packet loss rate increases: An increase in the number of data frames that get corrupted and dropped

QoS has the following elements, which are helpful in implementation and management of the network:

Traffic shaping and policing: Maintains efficient flow of traffic. This is done to prevent overflow of data packets at endpoints such as routers and switches.

Congestion avoidance: Monitors data traffic over the network to anticipate and avoid congestion.

Congestion management: Occurs in a network because of the presence of bandwidth-hungry applications such as multimedia applications using video and audio files.

QoS Architecture

The basic function of QoS is to provide better management and efficient flow of data information in a network. This is achieved through certain components and techniques that are a part of the QoS architecture. The main components of the architecture are as follows:

- QoS identification and marking technique facilitating QoS between all the endpoints in the network
- QoS within a single network with queuing, scheduling, and traffic-shaping tools
- QoS policy, management, and accounting functions to achieve smooth traffic flow in the network

Let us discuss all these components in detail:

QoS identification and marking: Marks the LAN segment or nodes that require preferential services. The second step is to mark the segment or the data packet to be given preference in the traffic flow process. Some of the common methods for identification are *Access Control Lists* (ACLs), *policy-based routing*, *Committed Access Rate* (CAR), and *Network-Based Application Recognition* (NBAR).

Congestion management: Includes managing congestion using certain tools. Congestion occurs when the amount of traffic passing through a network exceeds the speed of the link. The tools include:

First-In First-Out (FIFO): Has the store-and-forward capability, which means data packets will be stored in a buffer when the network is congested. When the network is free, this tool will forward the data packets in the order of arrival. Priority of individual packets is not considered with FIFO. It forwards each packet based solely on its arrival time.

Priority Queuing (PQ): Allocates priority to the data packets. When a data packet reaches the PQ, it is placed in one of the four queues: high, medium, normal, or low. During forwarding, packets marked with high priority are forwarded first and so on. PQ is configured statically and is not auto-configured.

Custom Queuing (CQ): Allocates appropriate bandwidth to the network users and applications. CQ is responsible for proportional distribution of the bandwidth. CQ is statically configured and does not have the auto-configuring feature.

Flow-based Weighted Fair Queuing (WFQ): Ensures bandwidth availability to the data traffic, including for low-priority traffic if there is no high-priority traffic in the queue. Unlike CQ, it is autoconfiguring in nature; that is, it is capable of adapting to network modifications or any other changes such as a node being added or deleted in the segment. WFQ uses *IP precedence bits* to provide priority services to the data packet in the queue. WFQ creates data packet flow based on the number of characteristics such as addresses (source as well as destination), socket numbers, and session identifiers. Each flow is then assigned its priority queue in case of congestion. When the congestion is clear, it forwards the packets.

Class-Based Weighted Fair Queuing (CBWFQ): Creates bandwidth slots for packets. Instead of creating flows as does Flow-WFQ, it creates a class that comprises one or more flow. Then, each class is assigned its priority.

Queue management: Concerned about congestion avoidance, that is, arranging data packets in such a way that it does not reach the traffic congestion point so that there are no network bottlenecks. The tools used are:

Weighted Random Early Detection (WRED): Includes algorithms to avoid congestion across internetworks. WRED monitors the traffic load on various points, and if it senses the increase in traffic, it discards or drops the traffic. The dropped packets are also referred to as *tail drops*. When the source detects the drop, it slows down the transmission. WREDs are designed for TCP/IP environments.

Flow RED: Used for non-TCP/IP environments in which the data packets are not dropped. Flow RED classifies the data into flows and then maintains the queue for forwarding the data packets. Flow RED also ensures that the packets in the queue do not take more time than is assigned to them.

Traffic shaping and policing: Used to maintain efficiency of the traffic flow by allocating proper bandwidth. It is used to avoid the problem of overflow of buffers in case of data packets. It limits the bandwidth flow of certain data packets to facilitate the flow of other packets based on priority. Shaping paces the data packets. If the traffic reaches or exceeds the configured rate, the data is buffered. Policing is similar to shaping. The only difference is that policing does not buffer the data packets but discards them. One of the tools used for implementing traffic shaping and policing is CAR. CAR is a policing tool used to limit the bandwidth of one flow to favor the bandwidth needed for a flow with higher priority.

Link efficiency mechanisms: Used to increase the efficiency of the traffic flow at the links or endpoints. Another benefit is that they eliminate the extra overhead bits that slow down the network. The mechanisms are implemented by these tools:

Link Fragmentation and Interleaving (LFI): Used to reduce delay and jitter in transmission of large data packets such as those originating from multimedia applications. It breaks up the large data packets and interleaves low-delay traffic packets with small packets. Due to LFI, delay-sensitive data does not wait for long after huge data packets.

Compressed Real Time Protocol (CRTP): Used for delay-sensitive traffic such as voice and video

Distributed Compressed Real Time Protocol (DCRTP): Allows more Voice over Internet Protocol (VoIP) traffic to pass through the network without decreasing the network performance. It also helps in reducing the size of the data packets, allowing higher transmission speed.

QoS management: Useful in setting and evaluating QoS policies and goals. It involves these steps:

- Baselining the network with utilities such as Remote Monitoring (RMON). This is useful in determining traffic characteristics of the network.
- Once the network characteristics have been evaluated, the next step is to deploy the QoS with the help of Cisco tools such as Cisco's *Quality of Service Policy Manager* (QPM) and *Quality of Service Device Manager* (QDM). For the purpose of verification at various service levels, deploy Cisco's *Internetwork Performance Monitor* (IPM).
- When the tools are deployed, the next step would be to evaluate the results and make regular improvements to achieve the QoS goals.

Table 6.2 sums up the features of QoS elements.

TABLE 6.2   Protocols Associated with QoS Elements

QoS Element	Protocols/ Tools Implementing the QoS Element
Traffic Shaping and Policing	• CAR • CAR Rate Limiting
Congestion Avoidance	• RED • WRED • Flow-based WRED • D-WRED
Congestion Management	• CBWFQ • Distributed LLQ • FIFO Queuing • IP RTP priority • LLQ • PQ • WFQ
Link Efficiency Mechanisms	MLPPP
Classification and Marking	• DCAR • MPLS EXP bit
Configuration and Monitoring	AutoQoS

SUMMARY

In this chapter, we learned about LAN segmentation and its advantages. We also reminded you about segmentation using different devices and the 80/20–20/80 rule. In the next chapter, we move on to designing Layer 3 addressing and naming conventions.

POINTS TO REMEMBER

- LAN segmentation reduces traffic congestion, increases bandwidth, enhances security, and isolates network problems.
- A LAN can be segmented through physical segmentation and network switching.
- Physical segmentation is done in three ways: using bridges, routers, and switches.
- *Transparent* and *source-route* are two types of bridging techniques.
- Spanning Tree Protocol (STP) is a link-management protocol that provides path redundancy for data transmission and eliminates looping in the network.
- Bridge Protocol Data Units (BPDUs) are packets containing information about the port addresses and priorities assigned to the frames.
- Routers are ideal devices to connect different LAN segments that use different media types.
- Different services provided by routers for a LAN segment are:

  IPX: SAP table updates

  AppleTalk: ZIP table updates

  Network Management: SNMP queries

  IP: Proxy ARP and ICMP

- Three types of methods used by switches for segmentation are *cut-through*, *store-and-forward*, and *fragment-free*.
- Switches provide higher port density at lower cost as compared to bridges.
- Microsegmentation is a concept of dividing a network into private or dedicated segments, that is, allocating one workstation per segment.
- Switches can maintain multiple simultaneous connections.
- A VLAN is segmented on a logical basis rather than on physical basis.
- Two types of VLANs are *static* and *dynamic*.
- Dynamic VLANs are not bound with physical ports as in the case of static VLANs.

- QoS helps avoid delay, lowered throughput, and excessive packet loss.
- The elements of QoS include *traffic shaping and policing, congestion avoidance,* and *congestion management.*
- Congestion-management tools include First-In First-Out (FIFO), Priority Queuing (PQ), Custom Queuing (CQ), Flow-based Weighted Fair Queuing (WFQ), and Class-Based Weighted Fair Queuing (CBWFQ).
- Queue management tools include Weighted Random Early Detection (WRED) and Flow RED.

7 Designing Third Layer Addressing and Naming Convention

IN THIS CHAPTER

- IP Addressing
- IP Management
- Address Management

IP ADDRESSING

Each technology has its own convention for addressing and transmitting messages between two machines in a network. For example, on a LAN, messages are sent between machines by providing a 6-byte unique identifier (the MAC address) to address the recipient of the message. Similarly, other network types, such as Token Ring and FDDI have their own addressing mechanisms. In addition to these specific network addresses, every workstation with TCP/IP installed is assigned a unique number. This number is called the IP address.

The Network layer of the OSI model provides a protocol-specific type of addressing known as *Network layer addressing* or *Layer 3 addressing*. It is also called logical addressing. Routers use this *logical address* to intelligently route data packets to the correct destination. For successful routing, the network is divided into several broadcast domains, and the packets are routed to each of these domains. This is done with the help of an IP addressing scheme.

The Hierarchical IP Addressing Scheme

In the hierarchical model scenario, the systems are structured in such a way that there is more than one cluster (group of nodes). Therefore, the hierarchical environment requires an IP address system, which will help in efficient routing of messages across the network. This is possible only when we have a well-defined addressing model. The hierarchical IP addressing scheme is capable of managing the IP addresses and routing tables in an expanding network.

A routing table contains information regarding routes to destination networks. The information in the routing table is updated by routing protocols.

IP Header Format

As already discussed, IP is a connectionless, unreliable datagram protocol responsible for addressing and routing packets between hosts. The header portion of an IP packet contains information to determine the characteristics of the packet, destination and source addresses, and protocol in use. Although the size of an IP header can vary, the limits are set. The reason for inconsistency in the IP header size is that not all fields are used during communication. The field size and other parameters of an IP packet are all set in the industry standard as laid down by the ISO. Therefore, hackers and crackers are able to sniff a packet on networks, and the packet information is easily available. Understanding the information contained in an IP packet will help to understand IP communication between two IP hosts. Figure 7.1 shows the format of an IP datagram header.

Table 7.1 briefly explains each field of an IP header.

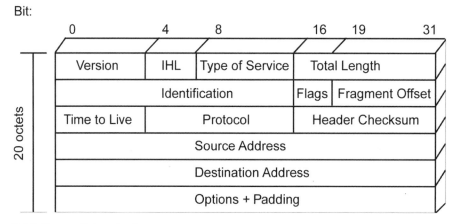

FIGURE 7.1   IP header format.

TABLE 7.1   Fields in an IP Header

Field	Description
Version (4 bits)	Identifies the IP version of the packet. It is a 4-bit field, which is usually set to binary 0100. *IPv4* is the current version in use on the Internet. The next version of IP addressing is *IPv6*, which is also referred to as *IPng* (IP next generation).
Internet Header Length (4 bits)	Identifies the length of the IP header. It specifies the length of the header in 32-bit words (4-byte blocks) and points to the beginning of the payload. The valid numbers range from 5 to 6, depending on whether additional "options" are used in the IP header. The size of IP header can range from 20 to 60 bytes in length (that is, the 4-bit header length field can have values from 5 to 15).
Type of Service (ToS) (8 bits)	Indicates the type of service required. These services are the levels of priority, delay, throughput, and reliability.
Total Length (16 bits)	Refers to the total length of the datagram including both the header and payload. Subtracting the header field from the total length field will determine the data payload of the packet. This field is measured in octets. The maximum size of a valid IP datagram is 65,535 bytes.
Identifier (16 bits)	Deals with the fragmentation of an IP packet. The value is set by the sender and is used for reassembling fragmented IP datagrams at the destination. Fragmentation of a packet is required when the packet's original MTU size exceeds the MTU size of the configured IP datagram on a data link through which it is passing. A device performing fragmentation first breaks an IP packet into smaller chunks and identifies them using this field.

(continued)

TABLE 7.1 (*continued*)

Field	Description
	Fragmented packets stay fragmented until they reach their final destination where they get reassembled
Flags (3 bits)	Comprised of three bits. Out of these three bits, the first bit is unused. The second bit is the Don't Fragment (DF) bit . If the DF bit is set to 1, that signifies that the packet cannot be fragmented. The last bit is the More Fragments (MF) bit. The MF bit is set to 1 on all fragmented packets except the last fragmented packet, indicating that the current packet is the last packet of the fragment.
Fragment Offset (13 bits)	Specifies the location of a fragment in IP packets. It is measured in 8 octet (64-bit) increments. The first fragment in the series always has an offset of 0.
Time-To-Live (TTL) (8 bits)	Specifies the maximum number of loops a packet can pass through to reach the destination host. The source computer sets the maximum TTL, then each router that handles the packet decrements the TTL by 1 and forwards the packet. If the TTL decrements to 0 before reaching the destination, the packet is discarded, and an error message is sent to the source. This process helps in dropping packets that would otherwise be subjected to endless looping in the network.
Protocol (8 bits)	Specifies the protocol to be used at the Transport layer of the DARPA *TCP/IP model*. For instance, if the datagram is a TCP packet, then this field is set to 6. If it is a UDP packet, this field would be set to 17. Other examples are: **Internet Control Messaging Protocol (ICMP): 1**

TABLE 7.1 *(continued)*

Field	Description
	Internet Group Management Protocol (IGMP): 2 **Interior Gateway Routing Protocol (IGRP): 88** **Open Shortest Path First (OSPF): 89**
Header Checksum (16 bits)	Performs a bit-level integrity check of the header information. It is basically an error correction field in the IP header and does not include the IP payload portion of the datagram. The source computer calculates the initial checksum before the packet is sent. As each router receives the packet, the checksum is verified, the TTL is decremented, a new checksum is recalculated, and then the packet is forwarded. If the checksum is incorrect at any router between the source and destination computer, the packet is discarded.
Source IP Address (32 bits)	Indicates the IP address of the source computer.
Destination IP Address (32 bits)	Indicates the IP address of the destination computer.
IP Options and Padding (variable)	This field is optional and can vary in length. If this field is used, it must be added in increments of 32 bits. That is, if the datagram has one option that takes up 8 bits, an additional 24 bits (all zeroes) must be added as padding. Therefore, the added option will add 32 bits (4 bytes) to the overall length of the IP header. This field is populated either by the source device or by the intermediary routers. The most frequently used options in the field are as follows: **Loose source routing:** In this case, the series of IP addresses to which the packet must be distributed are listed in the header. The packet must pass through

(continued)

TABLE 7.1 (*continued*)

Field	Description
	each of these addresses, although multiple hops may be taken between the listed IP addresses.
	Strict source routing: Like loose source routing, the series of IP addresses to which the packet must be distributed are listed in the header. Unlike loose source routing, however, the packet must not go to any destination address that is not on the list. If it does, an error is generated.
	Record route: Provides scope for each router to enter the address of its outgoing interface to a list in the "options" field in the IP header as the packet crosses it. This helps the source router keep a record of all the routers the packets encounter in their path.
	Time stamps: Provides options similar to a record route except that each router also enters a timestamp with the IP address. This keeps track of the packet.

IP Address Format

An IP network has two important resources, the IP addresses and the corresponding naming structure within the network. At the Internet layer of the DARPA *TCP/IP model*, all hosts must have a software address. With TCP/IP, the software address for a host is the Internet or IP address. This address identifies the network to which the host is attached. Routers use this address to forward messages to the correct destination. These addresses can be manually assigned to a host or can be dynamically assigned using a DHCP server. The process of assignment can be unmanageable for network administrators if done manually. The distribution and management of these addresses is an important consideration in an IP network design.

An IP address is a unique 32-bit Layer 3 address, which uniquely identifies a host and allows it to participate in a TCP/IP network. The IP address is divided into four groups called octets, each consisting of 1 byte (8 bits). This series is written in a dotted decimal notation with numbers ranging from 0 to 255. Table 7.2 shows a 32-bit IP address in a structured addressing scheme.

TABLE 7.2 A 32-bit IP Address

IP Address Notation	Address
Dotted decimal	192.168.12.1
Binary	11000000.10101000.00001100.00000001
Subnet mask (subnet masks will be discussed later in this chapter) in dotted decimal	255.255.255.0
Subnet mask in binary	11111111.11111111.11111111.00000000

As mentioned, the distribution and management of Network-layer addresses is important. Addresses for networks and subnets must be well planned, administered, and documented. As network and subnet addresses cannot be dynamically assigned, an unplanned or undocumented network would be difficult to debug and would affect the scalability of the network design. As opposed to the network itself, devices attached to the network can be configured for dynamic address allocation. This allows for easier administration.

The IP addressing scheme is set out in Request For Comments (RFC) 1166. The Internet address of a host contains both a network and a node address. Figure 7.2 depicts the IP address format consisting of the network number and host number.

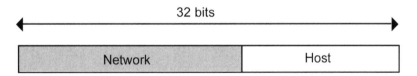

FIGURE 7.2 IP address format.

Network Number

For every computer, the network portion of its address must match the network address of other computers on that network. The length of the network part of the IP address depends on the class of the network being used. The network part of the IP address is important because routers use part of that information to route traffic from one network to another. These networks can use private IP ranges (set out in RFC 1918) as well as public IP ranges depending on the design and the requirement. If the network is internally managed under local administration, it can use the private IP range. If the requirement is to route traffic over a public network, then the network information/IP address needs to be obtained from the Internet Assigned Number Authority (IANA) (*www.iana.org*).

Host Number

The host portion of an IP address uniquely identifies the host in the network. For a given network part of the address, no two hosts can carry an identical host number if connected to the same network. Therefore, the host part has to be unique for that network number. This portion of the address is configured locally within the network by the authority that controls the network.

Each network can again be subdivided into a number of subnets, depending on the requirements. Finally, each host within the same network or subnet should have a unique IP address, which identifies it in the entire network. This is also called the *host address*. In Table 7.2, the host address is 192.168.12.1. The network part of the IP address identifies the network that the particular host belongs to. To communicate, all hosts should have the same network address. In Table 7.2, the network number is 192.168.12.0.

When a greater number of hosts and fewer number of networks are required, more bits are allocated to the hosts and fewer to the networks and vice versa.

For example, 192.168.2.4 is an IP address with a subnet mask of 255.255.0.0. The first two bytes or octets represent a network address, and the last two octets are the host address, as shown in Figure 7.3.

The number of bits assigned to the network and host portions depends on the number of networks to be configured. In a public network, the Internet Network Information Center (InterNIC) (*www.internic.net*) assigns this network number. To identify the number of bits that determine the network portion, a subnet mask has to be used along with the IP Address. The subnet mask determines the demarcation point between the network and the host portion of the IP Address. This is explained later in the chapter.

The IP address is assigned by the network administrator and differs from a MAC address, which is allocated by the hardware manufacturer.

NOTE

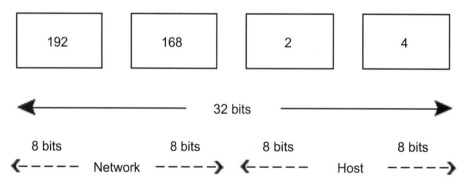

FIGURE 7.3 Allocation of IP addresses.

Classes of IP Addressing

The requirement to allocate IP addresses to networks of varying sizes was addressed by introducing the concept of *address classes*. The hierarchical model of the IP addressing scheme resulted in the use of different classes of IP addressing. The InterNIC assigns the classes of IP addresses to an internetwork with respect to the size of the network. This is to avoid any confusion during the allocation and distribution of IP addresses. There are five different address classes, A through E.

Internetworks are divided into three sizes:

Large internetworks: Assigned a larger number of hosts and fewer networks. These organizations are allocated Class A addresses.

Small internetworks: Assigned a smaller number of hosts and larger number of subnetworks. These organizations are allocated Class B addresses.

Medium internetworks: Assigned a requirement of hosts and subnetworks in between the big and small internetworks. These organizations are allocated Class C addresses.

Table 7.3 lists specifications and options associated with different classes.

TABLE 7.3   Class Requirement and Available Options

Class	Purpose	Maximum Number of Networks	Maximum Number of Hosts
A	Large organizations	127	16,777,214
B	Medium-sized organizations	16,384	65,543
C	Small organizations	2,097,152	254

The Internet community developed a set of rules in the hierarchical IP addressing scheme. For addresses in Class A, the leading bits of the first octet should always start with 0. The leading bits should be 10 for Class B, 110 for Class C, 1110 for Class D, and 1111 for Class E. Table 7.4 shows the address ranges of different classes.

TABLE 7.4   Leading Bits and Address Ranges of Classes

Class	Leading Bit	Address Range
A	0	1.0.0.0–127.255.255.255
B	10	128.1.0.0–191.254.0.0
C	110	192.0.0.0–223.255.255.255
D	1110	224.0.0.0–239.255.255.255
E	1111	240.0.0.0–254.255.255.255

Class A

Class A addresses range from 1.0.0.0 to 126.0.0.0, where the first octet represents the network portion, and the last three octets represent the host. The Class A address format is used for organizations with networks supporting a large number of users. The maximum number of networks possible with Class A addressing is 127, and the maximum number of hosts per network number is 16,777,214. The highest order of the network bits is always the most significant bit and determines the class of the network. In Class A networks, the highest order bit—the first bit of the first octet—is zero. Figure 7.4 depicts the Class A addressing format.

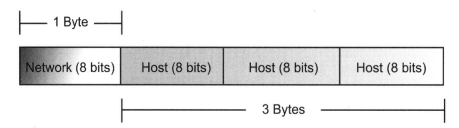

FIGURE 7.4   Class A addressing format.

Class B

Class B addresses range from 128.1.0.0 to 191.254.0.0, where the first two octets represent the network portion, and the last two octets represent the host. The Class B address format is used for networks of medium-sized organizations. The maximum number of networks possible with Class B addressing is 16,384, and the maximum number of hosts per network is 65,543. The highest order of the network bits is always 10. The first bit of the first octet is set to 1, and the second bit is set to 0. Figure 7.5 depicts the Class B addressing format.

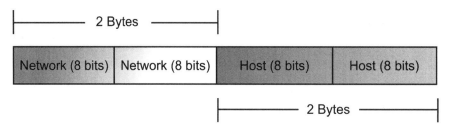

FIGURE 7.5   Class B addressing format.

Class C

Class C addresses range from 192.0.1.0 to 223.255.254.0, where the first three octets represent the network portion, and the last octet represents the host. The Class C addressing format is used for small organizations, with networks supporting a small number of users. The maximum number of networks possible with the Class C addressing is 2,097,152, and the maximum number of hosts per network number is 254. The highest order of the network bits is always 110. Figure 7.6 depicts the Class C addressing format.

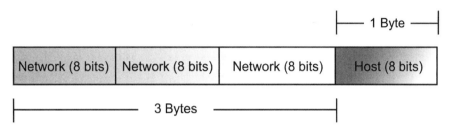

FIGURE 7.6 Class C addressing format.

Classes D and E

Unlike Classes A, B, and C, Classes D and E are not for commercial use. Class D addresses are used for multicast groups and range from 224.0.0.0 to 239.255.255.255. Class E addresses are used for experimental purposes and range from 240.0.0.0 and 254.255.255.255.

Table 7.5 shows a breakdown of IP addresses of Class A, B, and C in the binary format.

TABLE 7.5 IP Addresses in Binary Format

Network Number (Dotted Decimal)	Network Number (Binary)
10.1.1.0	00001010.00000001.00000001.00000000 (Class A)
150.5.5.0	10010110.00000101.00000101.00000000 (Class B)
192.1.1.0	11000000.00000001.00000001.00000000 (Class C)

In Table 7.4, note that Class A addresses have "0" in the first bit of the first octet, Class B addresses have "10" in the first two bits of the first octet, and Class C addresses have "110" as the first three bits of the first octet.

It is possible to connect networks with different Classes of IP addresses. Figure 7.7 depicts a scenario in which the networks 10.1.1.0 (Class A), 150.5.5.0 (Class B), and 192.1.1.0 (Class C) are internetworked.

FIGURE 7.7 Internetworking different classes of an IP Network.

In Figure 7.7, the networks belong to three different classes of address: 10.1.1.0/8 (Class A), 150.5.5.0/16 (Class B), and 192.1.1.0/26 (Class C).

Tables 7.6 and 7.7 list the characteristics of all five classes of IP addresses.

TABLE 7.6 Characteristics of Classes A, B, C, D, and E

Class	Format	Purpose	Leading Bit
A	N.H.H.H	Large organizations	0
B	N.N.H.H	Medium-sized organizations	10
C	N.N.N.H	Small organizations	110
D	N/A	Multicast groups	1110
E	N/A	Experimental	1111

TABLE 7.7 Characteristics of Classes A, B, C, D, and E

Class	Address Range	Maximum Networks	Maximum Hosts
A	1.0.0.0–126.0.0.0	127	16,777,214
B	128.1.0.0–191.254.0.0	16,384	65,543
C	192.0.1.0–223.255.254.0	2,097,152	254
D	224.0.0.0–239.255.255.255	N/A	N/A
E	240.0.0.0–254.255.255.255	N/A	N/A

There are some special function IP addresses, as follows:

Source address broadcast: When the host is in the initialization phase, all the bits from both the network and host parts of the IP address are set to zero (0.0.0.0). This is referred to as the source address. BOOTP and DHCP (discussed later in this chapter) are examples of such a scenario.

Destination address broadcast: When all 32 bits of an IP address are set to 1, it is referred to as a broadcast address. It is significant only when it is used as the destination address. When a message is sent with the destination address bits set to all 1s (255.255.255.255 in dotted decimal format), the message is broadcast in the segment to which the source is connected. As a result, all the hosts connected to the segment will receive the broadcast. Routers do not forward this kind of a local broadcast. There are two types of broadcasts in this case:

Directed broadcasts: In this case, only the host bits of an IP address are set to 1, and the network bits remain unchanged. This kind of a broadcast can travel to a destination crossing the routers. For example, in the IP address 192.168.1.255, all the hosts of 192.168.1 network will receive the broadcast.

Local broadcast: In this case, all the bits of an IP address (both the network and host parts) are set to 0. This type of broadcast is limited to the segment to which the source workstation is connected.

Loopback IP address: This is a special address set aside in the range 127.0.0.1-127.255.255.254. It is used to test the correct installation of the protocol stack of a host. Any message sent on this IP address will indicate the originating source.

Private IP address: When an organization has to route traffic in the Internet, it must have a valid IP address that is routable in the global Internet. If the organization does not have the requirement to route outside its own network, the network can be assigned any IP address that the local network administrator chooses. The range of IP addresses that are reserved (in RFC 1918) for private networks are given in Table 7.8.

TABLE 7.8 Range of Private IP Addresses

Class	Range of IP Addresses
A	10.0.0.0 to 10.255.255.255
B	172.16.0.0 to 172.31.255.255
C	192.168.0.0 to 192.168.255.255

NOTE

These IP address are not routable in the global Internet space, and therefore, can be used by any organization without reference to any other organization, including the Internet authorities. All external routers should discard any routing information regarding these addresses, and internal routers should restrict the advertisement of routes to these private IP addresses.

Public IP addresses are required by the organizations that need to communicate with the Internet. Any IP addresses that do not belong to the private IP range are all global/public IP addresses and can be used to route traffic in the public network.

The current and the most commonly used version of the TCP/IP protocol is IP version 4 (IPv4). However, it has some limitations, mainly the finite number of available addresses. Realizing that the number of addresses is being depleted, the Internet Engineering Task Force (IETF) is developing a newer version of the TCP/IP protocol suite called *IPng* or *IP version 6 (IPv6)*. IPv6 will offer stretched addressing capabilities with 128-bit addresses as compared to the 32-bit addressing supported by IPv4. Addresses from both IP versions will coexist as a backward compatibility measure.

Subnetting

Subnetting is a method of splitting a network into a collection of smaller networks called subnets. Subnetting is performed by borrowing host bits of an IP address because the network portion of an IP address cannot be changed (this process will be explained later). This may be done to reduce network traffic on each subnet or to make the internetwork more manageable. After subnetting, each of the subnet functions as an independent network. A host from one subnet communicates with the host of another subnet as if they were connected to a different network. For the hosts of two subnets to communicate, there is a need for routing. Without subnets, use of the network addressing space is inefficient.

An organization may break down the range of registered public IP address obtained from the InterNIC into several smaller internal networks, depending on the requirements. This calculation is based on the number of networks and hosts required by the organization. The subnet mask has an important role in implementing this. IP networks are subdivided according to the network design and requirements by using subnet masking. A subnet mask is a 32-bit address that indicates the number of bits of the IP address that represent the network address. The different subnets in a network are connected using a router. Subnetting efficiently uses subnet masks to provide flexibility for creating the requisite number of networks and the required numbers of hosts in a network, without wasting IP addresses. This decreases the number of broadcast domains, which in turn increases the performance of the network.

Subnets are locally created and administered within an organization's network. This means that the Internet community or users from other networks are unable to see the individual subnets within a network.

The binary representation of an IP address and its subnet mask performs a logical AND operation to find the network portion of the network. A logical AND operation is shown in Table 7.9.

TABLE 7.9 Logical AND Operation
Between A and B

A	B	A AND B
0	0	0
0	1	0
1	0	0
1	1	1

For example, a host with the IP address 132.16.12.129 has a subnet mask of 255.255.255.128. You can determine the network and the host portions by using a logical AND operation of IP address and subnet mask, as shown in Table 7.10.

TABLE 7.10 AND Operation of IP Address and Subnet Mask

IP Address Portion	Address
IP address in dotted decimal	132.16.12.129
IP address in binary format	10000100.00010000.00001100.10000001
Subnet mask in dotted decimal	255.255.255.128
Subnet mask in binary format	11111111.11111111.11111111.10000000
Logical AND between IP address and subnet mask gives the network portion	10000100.00010000.00001100.10000000
Derived network address in decimal format	132.16.12.128

Both the IP address and the subnet address are broken down into their binary format. The network address is derived by calculating the logical AND between these two values. Therefore, the network portion of the IP address 132.16.12.129 with the subnet mask 255.255.255.128 is 132.16.12.128.

As a part of subnetting, you borrow bits from the high-order bits of the host portion of the IP Address and use them to create the various subnets. The value 1 in a subnet mask specifies the network and 0 specifies the host.

Consider a major network 192.168.12.0/24 with the subnet mask 255.255.255.0. This network is divided into three different subnets of 192.168.12.128/28, 192.168.12.64/28, and 192.168.12.192/28. For an observer looking at the network from outside, say from the Internet cloud, the network 192.168.12.0/24 is the only one that is visible. The subnets are not visible. Figure 7.8 shows the subnetting of 192.168.12.0/24 into three different subnets of 192.168.12.128/28, 192.168.12.64/28, and 192.168.12.192/28.

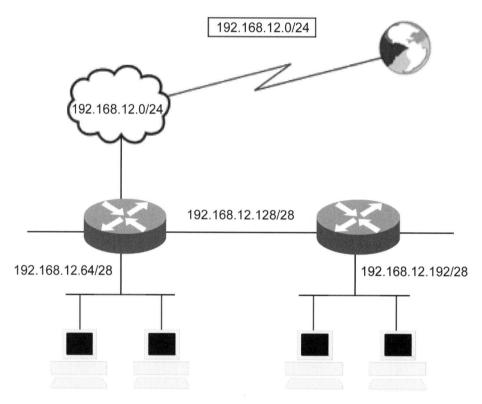

FIGURE 7.8   Subnetting 192.168.12.0/24 into three different subnets.

The subnet mask can also be written in the format '/N' where N is the number of 1s in the mask.

Subnetting is performed to create a smaller number of networks from a large network. A router interconnects these smaller networks. The router stops broadcasts, and the size of the broadcast domains becomes smaller. This reduces the network traffic on each subnetwork, increasing the performance. Small subnetworks are easier to manage, and the security policies can be fine-tuned to suit the requirement of each small subnet. The rules of subnetting are:

- Network bits cannot be all 0s or 1s.
- The boundary between the network and host portions of the IP address can be anywhere in the 32-bit range.
- The subnet mask determines the boundary between the network and the host portion of the IP address.

A Class A, B, or C network can be further divided or subnetted by a system administrator as per the network design or the requirements. After a network has been subnetted, the member hosts of those subnets reflect this subnet information. Now, the IP address carries a Network ID, unique Subnet ID in the network, and unique Host ID in the subnet, as shown in Figure 7.9.

Network ID	Subnet ID	Host ID

FIGURE 7.9   Subnetted IP address format.

This shows that when a network is divided into subnets, the host address portion of the IP address is divided into two parts, the subnet address and the host address.

For example, if a network has the Class B IP network address portion 172.16, the remainder of the IP address (host portion) can be divided into subnet addresses and host addresses. To differentiate the subnet ID and host ID from the host portion of the IP address, a subnet mask is used. This determines how the host portion of an IP address is divided into subnet address and the host address of that subnet. The bits of the subnet mask are set according to the following rules:

- All bits that correspond to the network ID are set to 1.
- All bits that correspond to the host ID are set to 0.

The process of dividing an IP network into smaller subnets involves two steps, which are as follows:

1. Determine the number of host bits to be used for subnetting. By knowing the number of host bits, you can determine the possible number of subnets and hosts per subnet. To finalize the number of host bits to be used, you need to know:

   - The number of subnets required
   - The number of hosts per subnet required

While determining the number of host bits to be used for subnetting, there are two major considerations to remember:

The more host bits used to form the subnet, the more subnets are created, but the number of hosts decrease: If you use more bits for the subnet mask than required, it will save you the time of reassigning IP addresses in the future. At the same time, it will waste some IP addresses. Creating subnets will lead to loss of IP addresses.

The lower the number of host bits used to form the subnet, the more hosts are created: This limits the growth in the number of subnets.

Two formulas that can aid in the calculation of the number of subnets and the number of hosts per subnet are as follows:

$$\text{Number of subnets} = 2^{\text{number-of-subnet-bits}} - 2$$
$$\text{Number of hosts per subnet} = 2^{\text{number-of-host-bits}} - 2$$

Table 7.11 shows the subnetting of a Class B network ID. Based on the required number of subnets and maximum number of hosts per subnet, a subnetting scheme can be chosen.

TABLE 7.11 Subnetting of Class B Network IDs

Number of Subnets Required	Number of Host Bits	Subnet Mask	Hosts Per Subnet
1–2	1	255.255.128.0 or /17	32,766
3–4	2	255.255.192.0 or /18	16,382
5–8	3	255.255.224.0 or /19	8,190
9–16	4	255.255.240.0 or /20	4,094
17–32	5	255.255.248.0 or /21	2,046
33–64	6	255.255.252.0 or /22	1,022
65–128	7	255.255.254.0 or /23	510
129–256	8	255.255.255.0 or /24	254
257–512	9	255.255.255.128 or /25	126
513–1,024	10	255.255.255.192 or /26	62
1,025–2,048	11	255.255.255.224 or /27	30
2,049–4,096	12	255.255.255.240 or /28	14
4,097–8,192	13	255.255.255.248 or /29	6
8,193–16,384	14	255.255.255.252 or /30	2

2. Derive the IP addresses and the new subnetted network IDs. In this step, calculate the IP address and the subnetted IDs depending on the number of host bits used for subnetting.

As an example, the subnetted IDs for the 3-bit subnetting of the private network ID 172.16.0.0 are shown in Table 7.12. The subnetted ID column displays the possible subnet ID while using 3 host bits for subnetting.

TABLE 7.12 Three-bit Subnetting for 172.16.0.0 Network

Subnet	Binary Representation	Subnetted ID
1	10101100.00010000.00000000.00000000	172.16.0.0/19
2	10101100.00010000.00100000.00000000	172.16.32.0/19
3	10101100.00010000.01000000.00000000	172.16.64.0/19
4	10101100.00010000.01100000.00000000	172.16.96.0/19
5	10101100.00010000.10000000.00000000	172.16.128.0/19
6	10101100.00010000.10100000.00000000	172.16.160.0/19
7	10101100.00010000.11000000.00000000	172.16.192.0/19
8	10101100.00010000.11100000.00000000	172.16.224.0/19

As an example, the range of IP addresses for the 3-bit subnetting of 172.16.0.0 is shown in Table 7.13.

TABLE 7.13 Range of IP Addresses for 172.16.0.0 Network

Subnet	Binary Representation	Range of IP Addresses
1	10101100.00010000.00000000.00000001– 10101100.00010000.00011111.11111110	172.16.0.1 to 172.16.31.254
2	10101100.00010000.00100000.00000001– 10101100.00010000.00111111.11111110	172.16.32.1 to 172.16.63.254
3	10101100.00010000.01000000.00000001– 10101100.00010000.01011111.11111110	172.16.64.1 to 172.16.95.254
4	10101100.00010000.01100000.00000001– 10101100.00010000.01111111.11111110	172.16.96.1 to 172.16.127.254
5	10101100.00010000.10000000.00000001– 10101100.00010000.10011111.11111110	172.16.128.1 to 172.16.159.254
6	10101100.00010000.10100000.00000001– 10101100.00010000.10111111.11111110	172.16.160.1 to 172.16.191.254

(continued)

TABLE 7.13 *(continued)*

Subnet	Binary Representation	Range of IP Addresses
7	10101100.00010000.11000000.00000001– 10101100.00010000.11011111.11111110	172.16.192.1 to 172.16.223.254
8	10101100.00010000.11100000.00000001– 10101100.00010000.11111111.11111110	172.16.224.1 to 172.16.255.254

Variable Length Subnet Mask

One of the original uses for subnetting was to subdivide a class-based network ID into a series of equal-size subnets. For example, a 3-bit subnetting of a class B network ID produced eight equal-sized subnets, using the all 1s and all 0s subnets. However, subnetting is a method of utilizing host bits to express subnets and may not require creation of equal-sized subnets.

Subnets of different sizes can exist within a class-based network ID. This is well suited to environments in which networks of an organization contain different numbers of hosts, and different-sized subnets are needed to minimize the wastage of IP addresses. The creation and deployment of various-sized subnets of a network ID is known as *variable length subnetting*, which uses a *Variable Length Subnet Mask* (VLSM).

VLSM is a method of dividing a network into different subnets by assigning different subnet masks. VLSM is designed to conserve IP addresses and optimize the use of the available address space.

Using VLSM, you can divide a network into granular subnets based on the requirements. In addition, you can allocate a subnet mask according to the number of networks and hosts.

The rules pertaining to subnetting using VLSM are as follows:

- Use a network address by allocating it to different hosts.
- Enable VLSM only for supporting routing protocols. This helps the routing protocol understand and carry the subnetting and supernetting (*supernetting* will be discussed later in the chapter) information required to route network traffic.
- For contiguous networks, summarize (defined later) so that the higher order bits are the same.
- Make routing decisions based on the longest matching network entry in the routing table.

Figure 7.10 depicts four remote routers connected to a central hub router. The LAN interfaces of the four routers have the following IP addresses: 172.16.1.65/30, 172.16.1.129/30, 172.16.1.193/30, and 172.16.1.225/30.

In the figure, after performing a logical AND operation between the IP address and the subnet mask, you will get the network address of each LAN interface. The derived network addresses are 172.16.1.64/30, 172.16.1.128/30, 172.16.1.192/30, and 172.16.1.224/30. This shows that a Class B network, which has the default subnet mask of 255.255.0.0, has a VLSM of 255.255.255.252. Using *route summarization*, the network address 172.16.1.0/24 is sent as a routing update. This step reduces the length of the routing table and efficiently utilizes bandwidth.

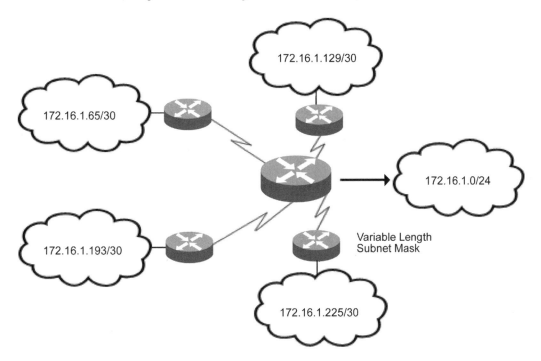

FIGURE 7.10 VLSM using four remote routers.

Route summarization is the consolidation of advertised subnetwork addresses such that a single summary route is advertised to other routers.

The advantages of VLSM are as follows:

- A VLSM ensures efficient use of the IP address range.
- A VLSM allows summarization, which reduces the length of the routing tables and saves bandwidth.

Protocols such as OSPF, Enhanced Interior Gateway Routing Protocol (EIGRP), Routing Information Protocol version 2 (RIPv2), and Border Gateway Protocol (BGP), support VLSM. Other protocols, such as RIPv1 and IGRP, do not support VLSM.

VLSM is different from CIDR because it is used within an organization. CIDR, which is discussed next, is used in the Internet.

Classless Inter-Domain Routing

The problems that have been encountered with IP address assignments have resulted in a move towards assigning multiple Class C addresses to organizations in preference to single Class B addresses. For each such Class C network, there must exist a separate routing entry in the routing table for these networks to route. The use of subnetting within a network can ease the addressability problems internally without placing undue burden on the routing tables of the external networks. Each external router will have to route each Class C address individually to your internal network. This method increases the routing overhead on the routers, and therefore, decreases efficiency. This problem has been named the *routing table explosion problem*, which is overcome by a technique called *Classless Inter-Domain Routing* (CIDR).

CIDR, or *prefix routing*, is a method to replace multiple routes by a single route entry in a top-level global routing table. IP addressing was not completely utilized when allocated to organizations that required a small number of hosts. For example, if an entire Class C address had to be allocated to an organization that required only 20 hosts, the remaining 234 (255-(20+1)) addresses were not used (the extra address is the broadcast address for that subnet).

Prefix routing was introduced to efficiently use the IP addressing space. Prefix routing and summarization have an important role in the Internet today because they provide a legitimate solution to the problem of the exhaustion of IP addresses.

If the prefix is shorter the network is more general and would represent larger number of subnetworks. If the prefix is longer, the network is a better match (i.e., more specific) and would include a lower number of subnetworks. At the top of the hierarchical design, the subnet masks get shorter because they aggregate the subnets that are lower down in the hierarchy. These summarized networks are also called *supernets*. The Internet uses supernets to aggregate different classes of addresses.

The suffix at the end of IP addresses (such as /24) is actually called a prefix, in Cisco. The prefix has a range of 1–32.

CIDR is set out in RFC 1817 and includes the process of aggregating multiple network numbers in a single routing entry. Route aggregation allows routers, which are at the top of the hierarchy, to group many routes into few routes. This aggregation saves bandwidth when the routing information is transferred to other routers. With

CIDR, several networks appear as a single entity to the outside network. Table 7.14 lists some examples of prefix masks.

TABLE 7.14   Examples of Prefix Masks

Prefix	Mask
/23	255.255.254.0
/27	255.255.255.224
/10	255.192.0.0
/6	252.0.0.0

CIDR allows flexibility of allocating networks and subnets. The optimization of routing table size and reduction of router resource overhead reduces the usage of network bandwidth.

For example, 192.168.48.0 is the network address of the hypothetical company Abbakis Inc. If it were the default subnet mask of a Class C network, it would have been 255.255.255.0. By using the subnet mask 255.255.248.0, the organization can use three additional rightmost bits of the third octet to get $2^3 = 8$ "extra" networks. Table 7.15 shows sample prefix masks to provide "extra" networks to an organization.

TABLE 7.15   Prefix Masks to Create Extra Networks

Description	First Octet	Second Octet	Third Octet	Fourth Octet
IP address (decimal)	192	168	48	0
IP address (binary)	11000000	10101000	00110000	00000000
Prefix mask (decimal)	255	255	248	0
Prefix mask (binary)	11111111	11111111	11111000	00000000

Abbakis Inc. is a large organization with a large network setup. The network setup consists of two routers connected over the Internet. One of the routers has networks 192.16.12.160/27 and 192.16.12.192/27, summarized to 192.16.12.128/25. The other router has networks 192.16.11.64/27 and 192.16.11.32/27, summarized to 192.16.11.0/25. Instead of sending the routing updates of each of the individual

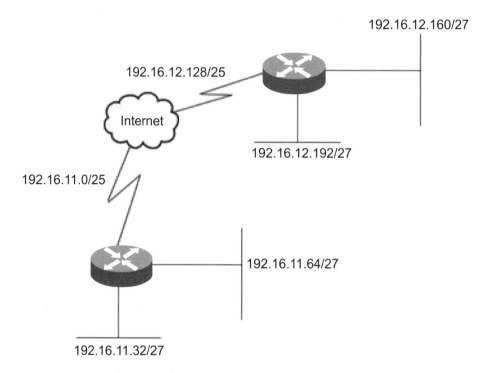

192.16.12.160/27

192.16.12.128/25

Internet

192.16.12.192/27

192.16.11.0/25

192.16.11.64/27

192.16.11.32/27

FIGURE 7.11 Prefix routing by sending only the summarized address.

subnets of the Internet, only the summarized address is sent, as shown in Figure 7.11 As a result, the use of prefix routing has helped preserve bandwidth.

Table 7.16 lists the prefix routing summarization for networks shown in Figure 7.11.

TABLE 7.16 Prefix Routing Summarization

Networks (Dotted decimal)	Networks (Binary)
192.16.12.160/27	11000000.00010000.00001100.10100000
192.16.12.192/27	11000000.00010000.00001100.11000000
192.16.12.128/25 (summarized)	11000000.00010000.00001100.10000000
192.16.11.64/27	11000000.00010000.00001011.01000000
192.16.11.32/27	11000000.00010000.00001011.00100000
192.16.11.0/25 (summarized)	11000000.00010000.00001011.00000000

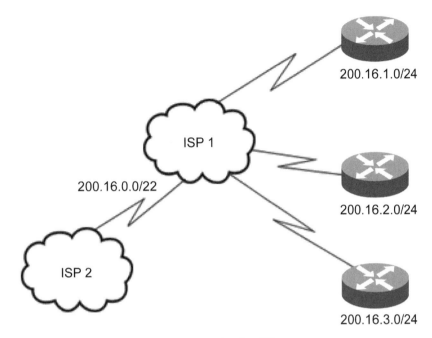

FIGURE 7.12 Prefix routing summarization by different ISPs.

For example, networks 200.16.1.0/24, 200.16.2.0/24, and 200.16.3.0/24 are networks that belong to different customers in an organization. These networks are summarized to 200.16.0.0/22 by ISP 1 and sent as routing updates to ISP 2. The ISPs conserve bandwidth by sending out prefixes during routing updates. Figure 7.12 shows the use of prefix routing summarization by different Internet Service Providers (ISPs) while exchanging routing updates.

To create supernets, you use different commands for each routing protocol. This will be explained in detail when discussing routing protocols. For example, the supernet command in EIGRP is:

```
Router (conf)# ip summary-address eigrp 1 200.16.0.0 255.255.252.0
```

IP MANAGEMENT

Management of IP addresses is a complex task. IP management involves managing issues such as assigning new IP addresses and mapping the existing addresses to the new ones. This includes managing the IP addresses, which are unused, and later assigning them to the system linked in the network. IP management helps in sorting out the network complexity with the help of certain tools, conventions, and methodologies, as discussed in this section.

Name Management

Due to the complexity in remembering the IP addresses of computers, it was decided that a translation mechanism would assist in assigning names to the IP addresses. These names are referred to as *host names*. In large networks, where remembering IP addresses is not feasible, host names are always preferred over IP addresses. Therefore, the need for an efficient name management solution was determined. The next section explains the name management solution, from the traditional methods to the new breed of technology available today.

Name Resolution

A host name is an alias assigned to an IP node to identify it as a TCP/IP host. The host name can be up to 255 characters long. Multiple host names can be assigned to the same host.

Host names have two components: the first component is the nickname and the second component is the *domain name*. The combination of these two makes up a complete host name, also referred to as a *Fully Qualified Domain Name* (FQDN). An FQDN uniquely identifies the host position within the DNS hierarchical tree by specifying a list of names separated by dots. A nickname is an alias to an IP address that individual people can assign and use. A domain name is a structured name that follows Internet conventions. It is discussed in detail in the DNS section.

Services such as HTTP and FTP can use either of the two methods to specify the destination to be connected—the IP address or a host name. When the IP address is specified, name resolution is not needed. When a host name is specified, the host name must be resolved to an IP address before starting an IP-based communication with the desired service.

Static Files

The simplest form of name resolution was achieved through the use of static files on each host system running or supporting TCP/IP. Information about name management using the static files is specified in RFCs 606, 810, and 952, of which RFC 952 is currently in force. These RFCs set out the use of hosts.txt and the method/structure with which an IP address can be matched with a host name.

The static file that contains this IP address-to-name mapping is called the *hosts file*. This file contains the listings of the host with other hosts' host names and associated IP address. By default, any host running the TCP/IP protocol suite creates this file in the directory path as listed here, in both Unix and Windows environments.

The directory path in a Unix environment is:

/etc/hosts

To update this file, manually edit or modify it with the information using any text editor.

The directory path in Windows environments is:

C:\WINDOWS\system32\drivers\etc\hosts

To update this file, manually edit or modify it with the information using any text editor.

To access the static file in the Cisco environment, while the router is in privileged mode, run the show host command. The output showing the mapping of the alias name ftp.test.com and www.test.com with their respective IP addresses is:

```
Router#show host
Default domain is not set
Name/address lookup uses static mappings
Host                  Flags       Age Type   Address(es)
ftp.test.com          (perm, OK)  0   IP     11.1.1.101
www.test.com          (perm, OK)  0   IP     11.1.1.100
```

To update this mapping information, use the command ip host [name] [IP address] in the global configuration mode of the router. This has to be done manually. An example of manually updating the IP hosts table in Cisco routers is:

```
Router>en
Router#conf t
Router(config)#ip host ftp.test.com 11.1.1.101
Router(config)#ip host www.test.com 11.1.1.100
```

If the hosts file exists in the system, it will contain a list of all the hosts that the system is required to communicate with. Similarly, all the systems that need to communicate using the host name should have this file updated with the assigned name and IP address of the device with which they intend to communicate.

You can manually update the file for each of the communicating devices, although this is the most inefficient method of maintaining the hosts files. This is because any addition of a new host to the existing network requires updating each device having this file. The size of the file is directly related to the number of hosts a system requires for name resolution. In very small networks, this method works well, but as the network increases in size, this method becomes unmanageable.

Another way of updating the file is by updating a centrally maintained static hosts file and then copying the file to each of the relevant workstations. This method can be considered as a substitute method for the manual method, and it will reduce the administrative overhead incurred by having an administrator visit each device. With both methods, administrators will have to intervene to maintain up-to-date lists of host names and IP addresses. In addition to this manual update of files, if the network administrator does not centrally manage the files, host name conflicts will occur and the size of the host file will become too large to be transferred across the network, especially over low-bandwidth WAN links.

Host Name

A host name is the left-most portion of a domain name that identifies a specific named host on the network. Figure 7.13 illustrates the host name of an IP host.

In Figure 7.13, the "abc" portion of the name refers to a specific IP host. The IP address assigned to these hosts can now be referred to by its alias "abc." The second part of the name refers to the domain name, that is, "test.com" is the domain portion of the address, indicating the domain in which the host resides.

The domain portion of the address may actually indicate multiple domains, including the root domain and one or more subdomains. In a complete host name, domain/subdomain is separated with a dot from the other domains/subdomains.

abc.test.com

Host   Domain Name

FIGURE 7.13 Representation of a host name.

Address Assignment

After the IP addresses for the network are obtained, the next step is to assign them to valid hosts. The IP address assignment to the systems across a network is managed by a network administrators/registrar. There are different methods for assigning IP addresses to hosts: manually by static assignment, and automatically by DHCP. The currently used forms of IP address assignment are static and dynamic.

Static Assignment

Static assignment indicates manual assignment of IP addresses to IP hosts. This mode of IP address assignment is done by network administrators. It involves assigning of IP network addresses by ISPs at the time of registration with the network. With the growth in size of many networks, it became difficult to manually assign static addresses to the systems. In addition, the static mode of address assignment has no support for diskless workstations, and this type of network is expensive to maintain. To overcome these problems, methods of dynamic IP address assignment were introduced.

Dynamic Assignment

Dynamic assignment allocates the IP addresses dynamically when a user plugs into the network. When a computer or other node joins the network, it is assigned a temporary IP address from a pool of IP addresses. This allows each node to maintain a unique identity. This automatic assigning is done with the help of the following protocols:

- Reverse Address Resolution Protocol (RARP)
- Bootstrap Protocol (BOOTP)
- Dynamic Host Configuration Protocol (DHCP)

Reverse Address Resolution Protocol (RARP)

RARP is a simple method suitable for diskless hosts on a small network. It determines the IP address from a host hardware address. This is just the reverse of ARP, which is used to determine a host's hardware address from the IP address. To use this method, install a RARP server. The RARP method hasn't been successful in large networks due to administrative overhead because the administrator needs to manually maintain the hardware address-to-IP address database in the RARP server.

Bootstrap Protocol (BOOTP)

BOOTP enables a client workstation to initialize with a minimal IP stack, and request its IP address, gateway address, and the address of a name server and other configured items from a BOOTP server. BOOTP is set out in RFC 951 and was designed to overcome the deficiencies of RARP. A good example of a client that requires this service is a diskless workstation. In this case, the host initializes a basic IP stack with no configuration to download the required boot code. Hosts with the local storage capability also use BOOTP to obtain their IP configuration data. However, the limitation of this protocol is that both the BOOTP server and client need to be on the same physical network (either an Ethernet segment or a Token Ring).

Dynamic Host Configuration Protocol (DHCP)

DHCP is the latest technology used in the automatic IP configuration of a host. It configures IP addresses and other items. DHCP is set out in RFC 951 and RFC 2132.

DHCP is based on BOOTP. This protocol adds the capability to automatically allocate reusable network addresses to workstations or hosts.

The DHCP standard provides the use of DHCP servers as a method to manage dynamic allocation of IP addresses and other related configuration details for DHCP-enabled clients on the network. Every node on a TCP/IP network must have a unique IP address. The IP address with its related subnet mask identifies the host computer and the subnet to which it is attached. When a workstation is moved to a different subnet, the IP address must be changed. DHCP allows you to dynamically assign an IP address to a client from a DHCP server IP address database on your local network. Figure 7.14 illustrates the concept of the DHCP server.

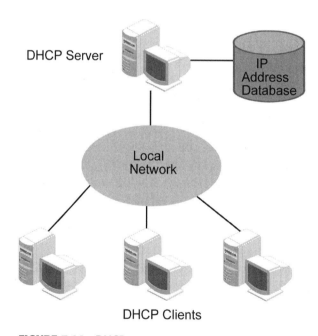

FIGURE 7.14 DHCP.

The DHCP service delivers the following benefits:

Safe and reliable configuration: Helps avoid configuration errors caused by manually typing values at each node. Also, DHCP prevents address conflicts caused by a previously assigned IP address being reused to configure a new workstation on the network. This is because a DHCP server leases only the unused address to the DHCP clients.

Reduces configuration management: Decreases time spent in configuring and reconfiguring computers on your network. Servers can be configured to supply a full range of additional configuration values while assigning address leases. These values are assigned using DHCP options such as default gateway, DNS servers, WINS server, and time server. Figure 7.15 shows a DHCP IP lease process in operation.

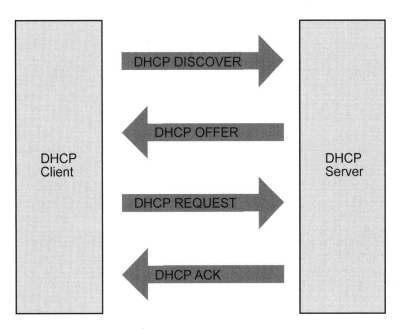

FIGURE 7.15 DHCP IP lease process.

When a new host configured as a DHCP client is initialized in the network, the following process takes place:

1. A broadcast message is sent to the network and received by the DHCP server(s). This broadcast message is known as the DHCP DISCOVER.
2. All available DHCP servers respond to the broadcast with their IP offer. This broadcast message is known as the DHCP OFFER.
3. The client selects the most appropriate server and sends a confirmation broadcast to the selected server. This broadcast message is known as the DHCP REQUEST.
4. When a server is selected, it sends the necessary configuration information to the client. This broadcast message is known as the DHCP ACK.

ADDRESS MANAGEMENT

All the nodes connected in a network are allocated unique IP addresses. This allocation is performed with the help of certain naming convention methods, which are discussed here.

Domain Name System

When the Internet was a small network, the host names of the computers in the network were managed through the use of a single HOSTS file located on a centrally administered server. Each site that needed to resolve host names on the network had to download this file. With the growth of the Internet, the number of hosts on the Internet grew. As a result, the traffic generated by the update process increased as well as the size of the HOSTS file. The need for a new system that would offer features such as scalability and decentralized administration became more and more obvious.

The presence of a Domain Name System (DNS) was needed to resolve the problems that arose with the increasing number of hosts. DNS was designed to have a naming scheme similar to a directory structure that organizations use. The primary objective of DNS was to resolve human readable addresses, such as abc.test.com into Internet IP addresses. In addition, DNS also helps to decentralize storage and management of the namespace, and provides lookup services. This technology provides more scalability compared to the static name resolution methods.

The root (the highest level) of the domain namespace is managed by the Inter-NIC. The InterNIC is also responsible for delegating administrative responsibility for portions of the domain namespace and registering domain names.

With DNS, the host names reside in a database that can be distributed among multiple servers, thereby decreasing the load on any one server. This enables administration of the naming system in a distributed manner. DNS supports hierarchical names and allows registration of various data types along with the host name-to-IP address mapping used in HOSTS files. The conceptual naming system on which DNS is based is a hierarchical and logical tree structure called the *domain namespace*. By virtue of the DNS database being distributed, its size is unlimited and performance does not degrade when adding more servers. Therefore, DNS provides an efficient scalability option.

DNS is an open standard, and there are various implementations of available DNS servers. Domain names are managed through the use of a distributed database system of the name information that is stored on name servers located throughout the network. An information store containing listings of alias names with their respective IP addresses is known as the domain namespace. Each name server has database files (known as zones) that contain recorded information for a selected region within the domain tree hierarchy.

The domain namespace is a hierarchical name database containing the host-name-to-IP address mapping in the Internet.

TIP

The domain name identifies a domain's position in the name tree relative to its parent domain. For using and administering a DNS service, the domain namespace refers to a domain name tree structure—from the top-level root to the bottom-level branches (subdomains) of the tree. To represent the host name in a DNS naming format, a naming convention is followed. According to this naming convention, for each domain level, a period (.) is used to separate each subdomain descendent from its parent-level domain. Figure 7.16 illustrates the DNS namespace for a company.

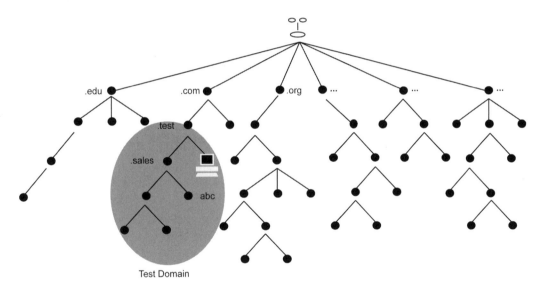

FIGURE 7.16   Hierarchical representation of abc.test.com.

In Figure 7.16, the "Root" domain and first-tier domains .net, .com, and .org represent the part of the Internet namespace, which is under administrative control of the Internet governing body (InterNIC). The second-tier domain Cisco, and its subdomains West, South, East, and Sales all represent the private namespace, which is under administrative control of the organization, Cisco. The FQDN of the host SRV1.Sales.Contoso.com determines exactly where this host resides in the namespace relative to the root of the namespace.

Architecture of Domain Namespace

The architecture of the domain namespace is split into multiple layers. The various layers of the namespace are as follows:

Root domain: Refers to the root node of the DNS tree architecture. It is represented as ".". (dot). It does not have a text label. It is sometimes represented in DNS names by a trailing period (.) to designate that the name is at the root or highest level of the domain hierarchy.

Top-level domain: Refers to the right-most portion of a domain name. A top-level domain is stated as a two-character or three-character name code that identifies either organizational or geographical status for the domain name. In the example abc.test.com, the top-level domain name is the ".com" portion of the domain name that indicates that this name has been registered to a business organization for commercial use (the .com code represents the commercial group).

Second-level domain: Refers to a unique name of varying length that is formally registered by InterNIC to an individual or organization connecting to the Internet. In the example of abc.test.com, the second-level name is the ".test" portion of the domain name, which is registered and assigned by InterNIC to the company Test (.test represents a company).

Subdomain name: Refers to branching the domain name into subdomains identified by separate name portions. A large organization may further subdivide its registered domain name into branches by adding subdivisions or departments that are identified by a separate name portion. For example, in abc.sales.test.com and xyz.finance.test.com, sales and finance are branching off the domain name "test" into subdomains.

DNS Server

A computer that is running the DNS service is considered a DNS server. A DNS server performs the following services:

- Stores namespace information for one or more domains
- Delegates parts of a domain to other servers
- Resolves names and addresses requested by clients for which it is authoritative

A zone is the portion of a domain's namespace for which a server has direct access, including pointers to delegate servers. Zones may be considered as subsets of domains.

NOTE

The DNS service resolves name resolution requests submitted by DNS clients (resolvers). DNS server consists of the following two elements:

Name server: Responds to browser requests by supplying name-to-address conversions

Resolver: Requests another name server for information when a name server does not have the result for the name resolution request

When an application needs to resolve a URL, it queries the closest name server. If that server has serviced a request for the same host name (within a time period set by the administrator to prevent passing old information), it will locate the information in its cache and reply. Otherwise, if the name server is unfamiliar with the domain name, the resolver will attempt to solve the problem by asking a server further up the tree. If this does not work, the second server will ask additional servers until it finds one that identifies the domain name. After the information is found, it is passed back to the client's application.

TIP

When a server supplies an answer without requesting another server, it is known as an authoritative server.

The terms associated with the storage of DNS data on a DNS server are as follows:

- Zone
- DNS Resource Records (RRs)
- DNS Queries
    - Recursive Queries
    - Iterative Queries

Zone

A *zone* is a portion of the DNS database that contains the resource records with the owner names that belong to the contiguous portion of the DNS namespace. Zone files are maintained on DNS servers. A single DNS server can be configured as a zone zero, zone one, or a server that contains multiple zones. Zones may store records in flat text files or in the Microsoft® Active Directory® database. Zones are classified into the following:

Forward lookup zone: Contains name-to-IP mappings

Reverse lookup zone: Contains IP-to-name mappings

DNS Resource Records (RRs)

DNS RRs are individual entries in a DNS database. Different record types represent different types of data stored within the DNS database. For example, an RR may represent a host on the network or a type of service available on a server. Clients query this data to obtain names and addresses to establish communication or to locate/determine what services are available (and where) on the network. There are different types of records as given in Table 7.17.

TABLE 7.17 Types of Records

Record	Description
A (Host)	Represents a computer or device on the network. It is one of the most commonly used DNS records that resolves a hostname to an IP address.
PTR (Pointer)	Used for finding the host name that corresponds to an IP address. It is the reverse of A-type records. The PTR is found only in the reverse lookup zone and resolves a hostname to IP address.
NS (Name Server)	Facilitates delegation by identifying DNS servers for each zone. They appear in all forward and reverse look-up zones. Whenever a DNS server needs to cross a zone, it refers to the name server resource records for the DNS servers in the target zone. This record resolves the domain name (same as the parent folder) to the host name.
SoA (Start of Authority)	Identifies a primary DNS name server of the zone as the best source of information for the data within the zone. It is the first record in a zone file.
SRV (Service Record)	Indicates the network services offered by the host. It resolves the service name to host name and port.
CNAME (Alias)	It is a host name that refers to another host name. It resolves host name to host name.
MX (Mail Exchanger)	Indicates the presence of an e-mail server. It resolves a host name to IP address.

DNS Queries

A DNS query refers to a request for a name resolution made to a DNS server. A client system may issue a query to a DNS server, which may then issue queries to other DNS servers in case the resolution fails in the first case.

In a typical query, a DNS client will submit a target host name to a DNS server to learn the IP Address associated with the requested host name. If the DNS server provides resolution, it passes it back to the client or directs the query to another DNS server.

The DNS client-side functionality can exist on DNS servers as well as on systems running applications that need the name resolution service. DNS clients and DNS servers both initiate queries for name resolution.

NOTE

Recursive Queries

Recursive queries are initiated either by a DNS client or a DNS server. The reply to a recursive request is either a positive or a negative response. When a client initiates a recursive query for information about a specified domain name, the name server responds either with the required information or with an error message. If the name server does not have authority over the domain name in the query, it sends its own queries to other name servers to find the answer. Therefore, a recursive query puts the burden of delivering a final answer on the queried server.

The DNS client side issues a recursive query to its configured name server. The name server side receives a recursive query and acts as a DNS client. It queries other name servers until it resolves the client's query, and responds to the client with the resolved address or a failure message.

Figure 7.17 shows the working of a recursive query.

Iterative Queries

Iterative queries are used between DNS servers to find answers on behalf of a client. Iterative queries have the burden of finding the final answer on the querying client (or DNS server) via recursion. Answers to iterative queries can be either a positive or negative answer or a referral to another server.

Iterative queries ask for the "final answer" or the "closer server." Iterative queries are used between servers during resolution of client requests. For example, lower-level servers will issue iterative queries to top-level servers to reduce the workload on top-level servers.

When a client makes an iterative query, the name server will return the information if it has the relevant data. If it does not, it will respond to the query with names and addresses of other name servers for the client to try next, rather than

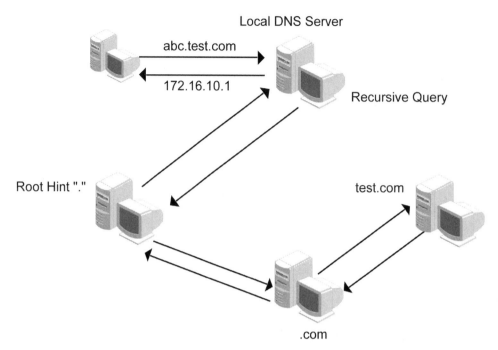

Local DNS Server

abc.test.com

172.16.10.1

Recursive Query

Root Hint "."

test.com

.com

FIGURE 7.17 Recursive queries.

asking other name servers for the data. A DNS server may respond differently for an iterative query depending on the situation. The list of responses that can be sent by the server includes:

- Requested address
- Authoritative "No"
- Referral (if the server recognizes the domain name being queried and knows a server address for that domain)
  Figure 7.18 shows the working of iterative queries.

Dynamic Domain Name System (DDNS)

While DNS is useful in providing IP addresses to the nodes in a network, DHCP is useful in distributing the IP addresses to the nodes. However, DHCP is not capable of updating the DNS server about the names and addresses it distributes. In that case, if new IP addresses are added, the network administrator/registrar needs to manually assign the IP addresses to the nodes. To overcome this limitation, the *Dynamic Domain Name Service* (DDNS) was introduced. DDNS is deployed with proper integration of a DHCP server and a DDNS-aware DNS server. A dynamic

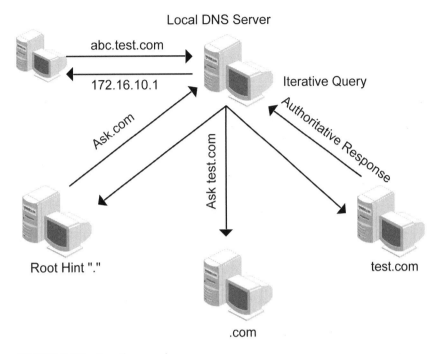

Local DNS Server

abc.test.com

172.16.10.1

Iterative Query

Ask.com

Authoritative Response

Ask test.com

Root Hint "."

.com

test.com

FIGURE 7.18 Iterative queries.

name server is capable of updating the lookup table whenever a DDNS-aware host or DHCP server informs the DDNS server to update a host name with the DHCP-allocated IP address. The DHCP server used in combination with a DDNS server eases the network and system administrators' workload from some time-consuming activities.

DNS, DHCP, and DDNS work in tandem to provide IP addresses to the organization or community of users by performing activities such as the following:

- DNS allocates the IP address to the organization or the user.
- DHCP manages the IP addresses.
- DDNS helps in automatically updating the DNS servers/database in case of new or modified IP addresses.

DDNS helps to automatically update IP names and addresses in the DNS database. This system brings in more automation by dynamically updating its resource record information about a new host or an existing host. It automatically updates the DHCP server about the new additions in the network, and DHCP updates the DNS server/database. A network registrar using DDNS can also inform the DNS server to remove an IP address from the database when the lease of the particular IP address expires.

SUMMARY

In this chapter, you learned about IP addressing scheme and fields related to it. We also reminded you about IP management, which included name and address management. In the next chapter, we move on to network management and troubleshooting.

POINTS TO REMEMBER

- The three layers of the Cisco hierarchical model are core, distribution, and access.
- Network layer addressing or Layer 3 addressing is also called logical addressing.
- The maximum size of a valid IP datagram is 65,535 bytes.
- The first fragment in the series always has an offset of 0.
- IP addressing is set out in RFC 1166.
- The private IP ranges are specified in RFC 1918.
- In a public network, the Internet Network Information Center (InterNIC) assigns the network number.
- The IP address is assigned by the network administrator. It differs from a MAC address, which is allocated by the hardware manufacturer.
- The maximum number of networks possible with Class A addressing is 127, and the maximum number of hosts per network number is 16,777,214.
- The maximum number of networks possible with Class B addressing is 16,384, and the maximum number of hosts per network is 65,543.
- The maximum number of networks possible with the Class C addressing is 2,097,152, and the maximum number of hosts per network number is 254.
- Class D addresses are used for multicast groups and range from 224.0.0.0 to 239.255.255.255.
- Class E addresses are used for experimental purposes and range from 240.0.0.0 to 254.255.255.255.
- Subnetting is done by borrowing host bits of an IP address because the network portion of an IP address cannot be changed.
- The Internet community or users from other networks are unable to see the individual subnets within a network.
- The process of dividing an IP network into smaller subnets involves determining the number of host bits to be used for subnetting and deriving the IP addresses and the new subnetted network IDs.

- Using VLSM, you can divide a network into granular subnets based on the organization's requirements, and allocate a subnet mask according to the number of networks and hosts.
- VLSM is used within an organization while CIDR is used within the Internet.
- The two components of host names are the nickname and the domain name, which together make up a complete host name, also referred to as a Fully Qualified Domain Name (FQDN).
- The command to view the manual mapping of the IP address with host names is show host. This command must be run when the router is in privileged mode.
- The domain portion of the address may actually indicate multiple domains, including the root domain and one or more subdomains.
- Automatic assigning is done with the help of RARP, BOOTP, and DHCP.
- The RARP method proves to be a failure in large networks because the administrator needs to manually map the hardware addresses to the IP address database in the RARP server.
- A DHCP server delivers safe and reliable configuration and reduces configuration management.
- DNS resolves human readable addresses into Internet IP addresses, helps to decentralize storage and management of the namespace, provides lookup services, and provides more scalability compared to the static name resolution.
- A DNS server consists of two elements, name server and resolver.
- A forward lookup zone contains name-to-IP mappings and a reverse lookup zone contains IP-to-name mappings.
- Two models of DNS queries are recursive and iterative.
- DDNS helps to automatically update IP names and addresses in the DNS database.

8 Selection of EMS and NMS

NETWORK MANAGEMENT

Network management is the process of managing a network; a LAN, WAN, or MAN, through a set of tools, procedures, protocols, and functions. Managing a network includes configuring and maintaining network operations, monitoring and controlling network performance, and diagnosing any network problem; that is, troubleshooting of the internetwork.

Need For Network Management Systems

In the early stages of development of networking environments, it was easy to configure and maintain a network because networks were rather small. With growing business needs and requirements, networks became more dynamic and started spanning several cities and countries. To manage such large networks, you need a network management system to help maintain and monitor the performance of the network across LAN and WAN connective devices.

Implementation of a Network Management System/Software (NMS) allows you to:

- Maximize the use of network devices for optimal output
- Easily identify and correct problem domains or potential problems in the network
- Provide remote notification and logging of network activity for analysis

- Minimize operating expenditures by reducing support staff needs
- Enable networks to achieve enhanced productivity

Network Management Architecture

All NMSs maintaining and monitoring a network follow a basic architecture that uses the same set of relationships. Network Management Architecture is based on the interaction of many entities with each other. The elements of this architecture are as follows:

Managed devices: These include hardware devices such as systems or nodes, printers, routers, or switches. They are also known as *network elements* and act as a platform for the management agent.

Management agents: These include software programs or modules that are located at the managed devices' end. They act as a connecting link between the managed device and the network management application. The management agents are responsible for compiling information related to managed devices, storing the information in the management database, and providing it to the management entities within NMSs through network management protocols.

Management Information Base (MIB): This refers to a collection of all managed objects grouped in a virtual information store. Managed objects in a network include modems and end-systems such as terminals, routers, switches, and others.

Network management application: This refers to an application that helps in monitoring and controlling the network. It prepares the log data in case of network discrepancies.

NMSs: This refers to devices that execute the network management applications. They are also known as *consoles*, and they include engineering workstations with fast CPUs and substantial memory so that they can execute commands easily and efficiently.

Management protocols: This includes protocols to transport management information between management agents and NMSs. Examples of management protocols are *Simple Network Management Protocol* (SNMP) and *Common Management Information Protocol* (CMIP).

Each network element in an internetwork runs agents (software modules) that send system alerts or error alerts in case they detect a problem in the network. After the problem has been detected, the management entities (management systems) execute certain steps such as operator notification, event logging, or system shutdown, to make it possible to repair the problem.

Figure 8.1 illustrates network management system architecture.

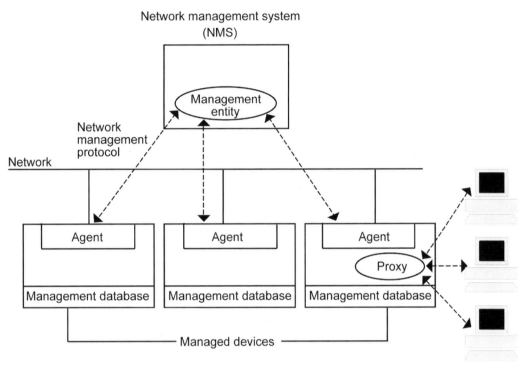

FIGURE 8.1   Network management system architecture showing protocols and agents.

ISO Network Management Model

The International Organization for Standardization (ISO) has developed a model for describing the functional areas of network management. This conceptual model helps in understanding the functionality of the NMS. The functional areas of an NMS are:

■ Fault management
■ Configuration management
■ Accounting management
■ Performance management
■ Security management

Fault Management

Fault management involves the following tasks:

1. Detection of errors
2. Maintenance of log reports

3. Isolation of the problem
4. Performance of diagnostics tests
5. Correction of the problem

When an error is detected, it is isolated. After isolating the network problem, perform diagnostic tests to fix the problem. Test the solution on all subsystems. Finally, make a record of the detection and resolution of the problem. Fault management mechanisms of the network model help the network administrators to detect or determine the problem in specific managed devices or endpoints across the network. For this, administrators can employ event collectors such as Syslog servers, and event producers such as SNMP and Remote Network Monitoring (RMON), in the network.

Configuration Management

Configuration management involves the following:

1. Initializing configurations
2. Reconfiguring network devices
3. Turning off managed devices

Configuration management is responsible for monitoring the network and system configuration information. This helps keep track of changes in the network related to hardware and software changes such as addition or removal of workstations, subnetworks, links, and other items, so that the changes do not affect the normal flow of information across networks. Configuration management subsystems store all the information in the form of databases for easy access. When a problem occurs, search this database for clues that may help solve the problem.

Accounting Management

Accounting management involves the following:

1. Providing billing information
2. Regulating groups and individual users
3. Maintaining the network performance at a standard level

Accounting management helps measure network utilization by users or groups. This allows for the minimization of network problems (because network resources can be allocated based on resource capacities) and maximizes the utilization of network resources equally amongst all users.

Performance Management

Performance management involves these tasks:

1. Gathering data based on network performance variables
2. Analyzing the data gathered in order to determine normal (baseline) levels
3. Determining performance parameters for each variable

Performance management helps in mapping and maintaining performance of the entire network with the help of certain performance variables such as network throughput, user response time, and line utilization. The performance thresholds are determined for each variable, so that if the managed devices exceed these parameters, an alert is generated and is forwarded to the network management system.

Security Management

Security management involves:

- Limiting user access to network resources
- Notifying administrators of security breaches and attempts

Security management helps control access to network resources to prevent compromise of sensitive information. In addition to monitoring user access, security management subsystems identify sensitive network resources and map them to user groups. They also log inappropriate access to sensitive network resources.

NETWORK MANAGEMENT INDUSTRY STANDARDS

After you have installed the network management system in your network, use network management protocols to successfully run the network management system. These protocols help network devices to communicate with each other and ensure effective network monitoring and control. This section discusses the following three network management industry standards:

- Simple Network Management Protocol (SNMP)
- Management Information Base (MIB)
- Remote Network Monitoring (RMON)

Simple Network Management Protocol

SNMP is a network management protocol used to monitor the functionality of various managed devices across the network, and also to enable information exchange between devices. It is an Application layer protocol and considered to be a part of the TCP/IP protocol suite. SNMP allows network administrators to monitor network performance, solve network problems, and increase the number of users in a network without seriously affecting performance. The three common versions of SNMP are SNMP v1, SNMP v2, and SNMP v3.

The Internet Engineering Task Force (IETF) developed SNMP in the 1980s by grouping RFC standards. Table 8.1 lists these standards.

TABLE 8.1   RFC Standards

Standard	Provides Information for
RFC 1155	Structure and Identification of Management Information for TCP/IP-based Internets
RFC 1157	SNMP
RFC 1212	MIB definitions
RFC 1213	Management Information Base for Network Management (MIB-II)

SNMP versions 2 and 3 included additional RFCs as listed in Table 8.2.

TABLE 8.2   RFCs Related to SNMP v2 and SNMP v3

Standard	Provides Information for
RFC 1901	Introduction to community-based SNMP v2
RFC 1902	Structure of Management Information of SNMP v2
RFC 1903	Textual Conventions for SNMP v2
RFC 1904	Conformance Statements for SNMP v2
RFC 1905	Protocol Information for SNMP v2
RFC 1906	Transport Mappings for SNMP v2
RFC 1907	MIB for SNMP v2
RFC 1908	Coexistence between Internet Standard Network Management Framework version 1 and version 2
RFC 2574	User-based Security Model (USM) for SNMP v3
RFC 2575	View-based Access Control Model (VACM) for SNMP v3

All Cisco equipment that supports IOS complies with SNMP. Both SNMP v2 and SNMP v3 provide better security and protocol management than SNMP v1.

SNMP Architecture

SNMP follows the network management model architecture and has the following elements to implement its services:

Managed devices or network elements: Consists of software called an SNMP agent. The network elements collect all the information pertaining to the management of the network, and in turn, transport the information to the NMS using SNMP.

Agent: This is a software module located in a network element or managed device. Along with the functions of a network element, an agent also translates the information collected by the element into a form compatible with SNMP.

NMS: Executes applications for monitoring and controlling managed devices.

Figure 8.2 depicts agents, the managed devices, and NMSs interacting with each other.

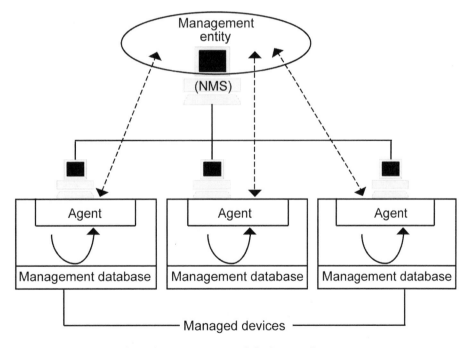

FIGURE 8.2 Interaction of agents, managed devices, and NMSs.

The network management stations execute management applications, which in turn, monitor the performance of an element for any problems or fixes. Thus the management stations control the entire network. For problem detection and solution, network elements have agents that implement SNMP for establishing communication between two endpoints—managed devices and network management stations.

SNMP Operations

SNMP performs two major functions: monitoring and controlling nodes such as workstations, routers, switches, and other network devices through a set of standard commands to enhance the performance of the network. Table 8.3 lists four basic SNMP commands that are used in operations:

TABLE 8.3   Basic SNMP Commands

SNMP Command	Used by	Function
Read	NMS	Monitors managed devices by examining their variables.
Write	NMS	Controls managed devices by changing the values of variables stored in managed devices in case of discrepancies.
Trap	Managed devices	Sends event logs to the NMS in case of an event occurrence such as a change in the status of the interface (up or down).
Traversal operations	NMS	Sequentially collects information in the form of tables such as routing tables. The information is extracted from the managed device by determining the variable type it supports. This table assists in the smooth flow of information and delivers high performance in a network.

In an SNMP network, communication is in the form of requests and responses. SNMP packets contain two parts. These are:

- Message header
- Protocol Data Units (PDUs)

Figure 8.3 illustrates a basic SNMP message format.

FIGURE 8.3 An SNMP message format showing message header and PDU.

Message Header

The message header contains two fields:

Version number: Specifies the SNMP version in use.

Community name: Creates an access environment for the group of NMSs. Community name acts as a form of authentication for devices in an NMS.

Protocol Data Unit (PDUs)

PDUs refer to the requests and responses sent by SNMP devices and transport and exchange information between the manager and agents.
The PDU contains five fields. These are:

PDU type: Specifies the type of PDU transmitted.

Request ID: Associates SNMP requests with responses.

Error status: Indicates errors and their types. This field is populated only by the response operation of SNMP packets. Other operations give the value 0 to the field.

Error index: Maps an error with a particular object instance. The response part of the SNMP packet sets this field. Other fields set its value to 0.

Variable binding: Associates an object instance with its current value. It is used as a data field of the PDU.

Figure 8.4 illustrates a PDU format.

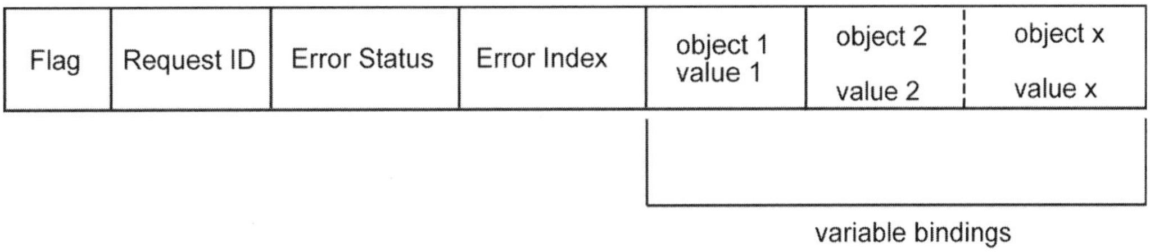

Flag	Request ID	Error Status	Error Index	object 1 value 1	object 2 value 2	object x value x

variable bindings

FIGURE 8.4 A PDU format showing various fields.

The structure of an SNMP trap PDU (a message or event generated in case of any malfunctions in the network) consists of these fields:

Enterprise: Identifies the type of managed object generating the trap

Agent address: Provides the address of the managed device that sent the trap

Generic trap type: Indicates the number of generic trap types

Specific trap code: Indicates a number of trap codes

Time stamp: Evaluates the time difference between the last network initialization and the trap generated

Variable binding: Refers to the data field of the trap PDU

Figure 8.5 illustrates the format of an SNMP trap PDU.

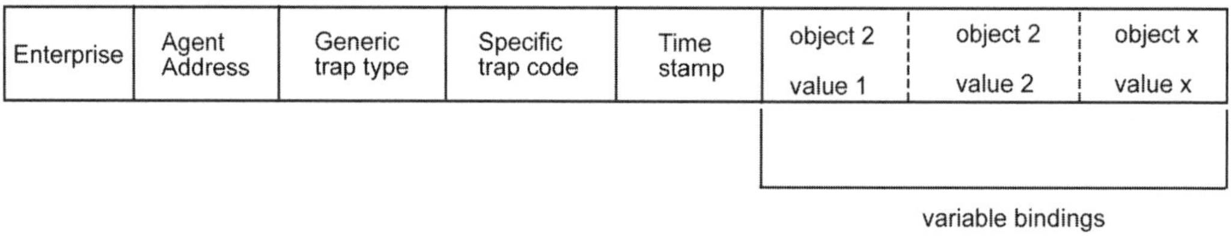

Enterprise	Agent Address	Generic trap type	Specific trap code	Time stamp	object 2 value 1	object 2 value 2	object x value x

variable bindings

FIGURE 8.5 An SNMP trap showing different fields.

Table 8.4 lists frequently used PDUs.

TABLE 8.4 Frequently Used PDUs

Command	Interpretation
Get	Allows NMS to retrieve an object instance (the value of the SNMP variable) from the table
GetNext	Allows the NMS to retrieve the next-in-line object instance from the table or the agent.
GetBulk	Used in case of SNMPv2 to acquire a large bulk of data
Set	Used by NMS to set values for object instances within the agent
Trap	Alert sent by the agent in case of an unusual event

SNMP v1 Operations

SNMP v1 has evolved from the basic SNMP protocol. The details about SNMP v1 are given in RFC 1157. Like SNMP, it is a simple request-and-response protocol and uses the same set of commands, including Get, GetNext, Set, and Trap.

Executing the Get command fetches the value of managed objects. GetNext obtains the value of the next managed object from the table contained within the agent. The Trap command is used to capture any event and inform the NMS.

The message format used in SNMP v1 is the same as described in the case of basic SNMP operations. The message format comprises the message header and PDU. The functionality of the message header and PDU is the same as explained earlier.

SNMP v2 Operations

SNMP v2 is an enhancement for SNMP v1. The basic operations and commands are similar with some additions. The commands are as follows:

Get bulk request: Allows an agent to respond to the manager (management station) with information in bulk

Inform request: Allows NMS stations (consoles where NMSs are deployed) to share trap information across the network

The SNMP v2 message format is similar to SNMP v1. The difference is that SNMP v2 has two PDU formats:

- PDU format
- GetBulk format

The PDU format is same as the SNMP v1 PDU format. The GetBulk PDU format, however, has these fields:

PDU type: Identifies the PDU as the GetBulk operation.

Request ID: Maps the SNMP request with the responses.

Non-repeater: Indicates the total number of object instances in the variable bindings field. Use this field when the object instance is scalar with only one variable.

Max repetitions: Sets the maximum number of times a variable should be retrieved, apart from that specified by nonrepeaters.

Variable bindings: Used as the data field of an SNMP v2 PDU.

SNMP v3 is an enhancement of both SNMP v1 and SNMP v2. The additional commands used in this version are:

No auth no priv: Authenticates authorized users based on the user name

Auth no priv: Authenticates the user based on the HMAC-MD5 or HMAC-SHA hashing algorithms

Auth priv: Authenticates the user and encrypts the data with the CBC-DES-56 algorithm (industry standard that helps in data encryption)

The NMS sends a periodic request for the status of the managed device using the GetRequest command, and the agent responds using the GetResponse command. The system of periodic requests is called "polling." The manager (NMS) is informed after a period of time about the status of the managed devices on the network. Therefore, any problem can be scaled easily. In case of an unusual event, the agent sends a trap to the manager. Some common trap commands are listed in Table 8.5.

TABLE 8.5 Common Trap Commands

Command	Interpretation
ColdStart	The agent has been restarted; the configuration may have changed.
WarmStart	The agent has been reinitialized without any configuration changes.
EnterpriseSpecific	An event specific to hardware or software
AuthenticationFailure	The agent received an unauthenticated message.
LinkDown	The agent detected a network link failure.
LinkUp	The agent detected a link status changing to up.
EgpNeighborloss	The device's EGP (Exterior Gateway Protocol) neighbor is down.

Management Information Base

MIB is the collection of information arranged in hierarchical order in the form of a database. MIB contains descriptions of all network objects (managed objects) identified by individual object identifiers. In addition, the managed objects contain one or more instance of each object, referred to as *variables*. MIB is accessed by SNMP protocols.

The two important constituents of MIB are:

Managed objects: This can be further categorized into:

Scalar objects: Single instances (variables) of an object

Tabular objects: Multiple instances of objects arranged in the MIB tables

Object identifier: Uniquely identifies a managed object in the MIB database

The hierarchical structure of MIB can be summarized as a tree with a nameless root. The levels of the tree nodes are assigned by different organizations. The top nodes are occupied by the standard organizations CCIT, ISO and ISO-CCIT. The lower nodes are assigned by the associated organizations. Private vendors have the discretion to create private branches for their own products. Figure 8.6 shows the structure of MIB.

PDU Type	Request ID	NON repeaters	Max repetitions	object 1 value 1	object 2 value 2

FIGURE 8.6   MIB structure showing different network objects.

Cisco maintains its MIB definition under the Cisco MIB subtree (1.3.6.1.4.1.9).

NOTE

Remote Network Monitoring (RMON)

RMON is a network management tool that enables various network devices to exchange network-monitoring data. RMON helps the network administrators to monitor, troubleshoot, and analyze any large or small network. The main purpose of developing RMON was to monitor and manage the LAN segments and other remote sites from a central location.

The user community created RMON with the help of the IETF. It was standardized in the year 1992 and was declared a draft standard in the year 1995. It was specified as a part of the MIB in RFC 1757 and considered as an extension of the SNMP protocol. Figure 8.7 depicts the RMON model.

RMON Components

RMON consists of the following two important components:

RMON probe/agent: Refers to a remote intelligent device or software agent that collects network status or statistics about a LAN or its segments and transfers the collected information to the management station.

RMON client referred as a management station: Refers to a device that coordinates with the RMON probe and collects statistics from it. The network administrators use it to monitor the network.

The RMON agent stores all the information related to the network in the RMON MIB. The MIBs are located or embedded in hardware devices such as

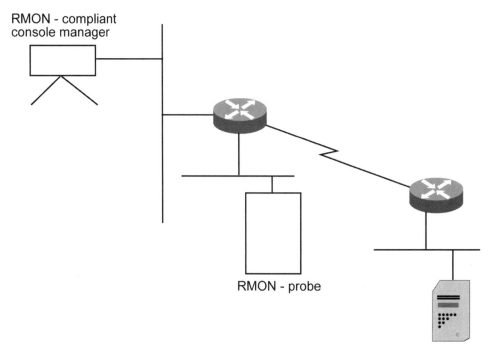

FIGURE 8.7 RMON monitoring a network.

routers and switches. The agents scan the network traffic that passes through them. They are located on each network segment. The management station, also called the *client*, communicates with the RMON agent to collect and correlate the network data with the help of SNMP.

RMON passes the network monitoring information to the network administrators in the form of RMON groups. There are nine subdivisions, each providing a specific set of data useful in managing and monitoring the network.

Table 8.6 lists the information details provided by the nine RMON groups.

TABLE 8.6 RMON Group Functions and Components

RMON Group	Functionality	Components
Statistics	Contains statistics probed by the agent for each device present on the network.	All data about the packets dropped and sent, and the number of broadcasts, CRC errors, runts (miniframes of less than 64 bytes), and collisions.

TABLE 8.6 *(continued)*

RMON Group	Functionality	Components
History	Contains recorded periodic samples from the network devices and stores them for later-stage retrieval.	Sample periods and item samples.
Alarm	Takes statistical samples from variables and compares them with previous configured thresholds. In case of discrepancies, it generates an alarm.	Contains the alarm table containing information such as alarm type, interval, and start and stop thresholds.
HOST	Contains the statistics connected with each host present on the network.	Contains the host address, data packets, bytes received, bytes transmitted, and broadcast and error packets.
HostTopN	Prepares a table describing the hosts. The statistics used in the table are rate-based.	Statistics, host, sample start and stop periods, rate base, and duration.
Matrix	Stores statistics used for communicating between two sets of addresses in the form of a table.	Source and destination address pairs and packets, bytes, and errors for the packets.
Filters	Provides specifications for matching packets with the filter equation. The packets form a stream that is monitored for event generation.	Bit-filter type, filter expression, and conditional expression to other filters.
Packet capture	Captures packets when they flow through a channel.	Size of the buffer and the number of captured packets.
Events	Controls the generation and notification of events from the devices on the network.	Event type, event description, last time an event notification was sent.

RMON is designed primarily to monitor and diagnose the network traffic contained in the MAC sublayer of the Data-link layer. Currently, there are two versions of RMON, RMON 1 and RMON 2:

RMON 1

RMON 1 was designed by IETF for monitoring and protocol analysis of Ethernet and Token Ring LANs. The key features of RMON 1 are network fault diagnosis, planning, and performance management.

RMON 1 provides the following features:

- Compilation of traffic statistics, both previous and current, for a network segment
- Alarm generation system for notifying network administrators of any abnormal network behavior
- Filter and capture facility used to build a complete distributed protocol analyzer

RMON 1 addresses network management issues pertaining to the lower layers, and is used for troubleshooting networks. It is not equipped to handle network management issues relating to higher layers. To address issues related to the higher layers, RMON 2 was developed.

RMON 2

RMON 2 is capable of monitoring Application-layer data traffic. However, it is not equipped to provide support to high-speed LAN/WAN topologies. The enhancements of RMON 2 over RMON 1 are as follows:

- Monitors Layer 3 traffic with the help of the Network-layer host and matrix tables
- Monitors Layer 7 traffic with the help of the Application-layer host and matrix tables
- Maps network addresses for the purpose of aggregating network statistics
- Contains protocol directory and distribution groups for the purpose of displaying selected protocols
- Contains samples of user-specified variables

Being a subset of SNMP, RMON provides wire management statistics and facilities. It is approved as an international standard and can be implemented over a wide range of LAN segments. However, it suffers from certain disadvantages. These are as follows:

- They are not equipped to provide sufficient information to network administrators for solving complex problems at remote locations.

- Slow data retrieval mechanism to a central management console.
- Inefficient bandwidth provision.
- RMON values stored in the 32-bit registers reduce the count value (maximum number that can be counted) to 4,294,967,295. For example, in a 100-Mbps Fast Ethernet network running at just 10 percent of capacity, the counters will reset to zero after just one hour of activity.
- A network in full compliance to RMON standards cannot be achieved by vendors due to the need for as yet nonexistent RISC processor technology that would involve huge manufacturing costs.

NETWORK TROUBLESHOOTING

Network management is all about documenting network behavior and understanding how the network operates under normal conditions. As a network administrator, you must also think about problems that may arise so that you can solve them more effectively. Being proactive, you can avoid serious scenarios such as network failures, which result in huge money and data loss to the organization. While managing networks, perform the following tasks:

- Collect network information and compare the collected statistics
- Perform routine tests and check for network health
- Design and specify goals for network management such as acceptable downtime, response time, and throughput levels
- Determine the minimum QoS for efficient network management

In case a problem has already been diagnosed in the network causing disruption in the normal flow of the data traffic, you can follow a plan of action to rectify it. The steps or strategies to be followed for safeguarding the network are:

- Developing metrics for measuring network performance
- Implementing network management systems to ensure sound network health
- Analyzing the data and documenting the results
- Locating network snags and bottlenecks
- Planning techniques to overcome problems of network failures

Network Baselining

Baselining refers to maintaining a record of network settings and configurations over a period of time to compare the current statistics with the previous collected statistics. Network baselining is used to assess the network health and performance and is an efficient technique to troubleshoot network problems. Network baselining involves a collection of various network and systems documentation, and includes collected information on the following items:

Network configuration table: Keeps records of all the hardware and software devices and components used in the network infrastructure

Network topology diagram: Provides a pictorial depiction of the network showing connectivity among various devices. This diagram also shows the direction of flow of data traffic.

End-system network configuration tables: Contains information about the items such as system name, system purpose, OS, VLAN, IP address, subnet mask, default gateway, and DNS server address(es).

End-system network topology diagram: Provides a graphical representation of the end-systems in the network for the purpose of giving a better view about the traffic flow.

Troubleshooting Approaches

Network troubleshooting requires a systematic approach because a small error in the troubleshooting process may lead to significant financial and productivity losses. To provide a systematic methodology to troubleshoot network problems, the three approaches to be followed are:

- Cisco Hierarchical Approach
- Layered Troubleshooting Approach
- Problem-solving Approach

Cisco Hierarchical Approach

Cisco uses the hierarchical model to design and troubleshoot a network. This model provides an efficient design topology to divide the network into three distinct layers: core, access, and distributed. These layers represent logical and not physical segmentation of the network. This way of segmenting the network reduces the complexity of internetworks consisting of multiple networks, communication subnets, and routers. This makes it possible to maintain each part of the network without disturbing the entire network.

Using the hierarchical model, you can design, maintain, and troubleshoot a network with respect to the following:

- Number of network users
- Number of running applications

■ Number of subnetworks involved in creating a complex network
■ Number of servers in the network and the estimated load that each server can effectively handle

Figure 8.8 displays the Cisco hierarchical model.

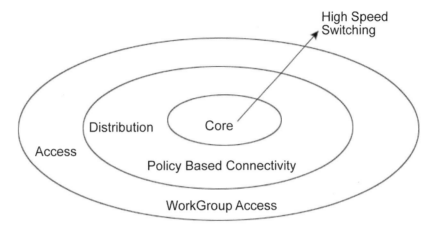

FIGURE 8.8   The Cisco Hierarchical Model showing different layers.

Let us learn about troubleshooting in each layer.

The Core Layer

The *core layer* is a high-speed switching backbone transporting large volumes of traffic quickly and reliably. The traffic transported across the core is between users and enterprise services, such as e-mails, videoconferencing, and dial-up access to the network. The links in the core layer are point-to-point.

NOTE

The services provided by the network are common to all the users of the network and are called enterprise services.

As the core layer is involved in high-speed transport, there is no room for latency and complex routing decisions pertaining to filters and access lists. Therefore, protocols such as OSPF and BGP, which have fast convergence time, are implemented at this layer. QoS may be implemented at this layer to ensure higher priority to traffic that may otherwise be lost or delayed in congestion. The core layer should have a high degree of redundancy.

This is the most important layer of a complex network and needs to be equipped with extensive troubleshooting support. This is because if there is a malfunction in the backbone, the entire network is rendered nonfunctional.

The Access Layer

The *access layer* enables user interaction with the network. It is also called the *desktop layer*. The user workstations and local resources such as printers are at this layer. Routers serve as gatekeepers at the entry and exit of this layer and ensure that the local server traffic is not forwarded to a wider network. The access layer controls the user and workgroup access to the internetwork resources. Other functions performed on this layer include sharing and switching of bandwidth, MAC-sublayer filtering, and microsegmentation.

Therefore, troubleshooting the access layer would require identifying problems of the local servers or network resources. In addition, problems at the access layer can include switch malfunction. This problem can be ameliorated by having backup switches available.

The Distribution Layer

The third layer in the Cisco hierarchical model is the *distribution layer*. This layer enables routing of data to the destination nodes and establishes WAN links between subnetworks. In addition, this layer performs data filtering functions to ensure security of data and the network. It is also known as the *workgroup layer*.

The distribution layer serves as a bridge between the access layer and the core services of the network. This layer examines the data packets and user requests to be allowed access to core services of the network with respect to the associated user request. The distribution layer also routes data packets through the fastest route to access the required core service of the network or the required destination node.

This layer implements network policies and controls network traffic and data movement. In addition, it performs complex CPU-intensive calculations pertaining to routing, filtering, inter-VLAN routing, Access Control Lists (ACLs), address- or area-aggregation, and security, and identifies alternate paths to access the core.

To prevent network congestion, the distribution layer also segregates the network into domains to distribute the load effectively over the entire network. In addition, the distribution layer also ensures compatibility among different types of networks, such as Ethernet and Token Ring.

Problems at the distribution layer are related to network congestion and non-functioning WAN links between the subnetworks. Troubleshooting the problems of the distribution layer involves restoring the WAN link and using appropriate filtering mechanisms or an appropriate security mechanism.

The Layered Troubleshooting Approach

As already learned, networks are based on the Open System Interconnection (OSI) model of networks—a layered architecture that is used to design networks compatible with all types of operating systems. The OSI model contains seven layers, which are organized in the order of their role for facilitating data transfer. In addition to the OSI model, another model that is widely used as a standard for internetwork communication is the Transmission Control Protocol/Internet Protocol (TCP/IP) model. This model has a four-layered architecture, and each layer corresponds to one or more layers of the OSI model. The functions of these layers are similar to those performed by the layers of the OSI model.

As most networks are based on the layered architecture of OSI or the TCP/IP model, the troubleshooting approach applied to such networks is called the *layered troubleshooting approach*. Using the layered approach, you can isolate and troubleshoot the problem pertaining to a specific layer. As a result, other layers are not affected, and troubleshooting is concentrated on the area of the network where the problem has occurred. The rest of the network functions without any interruption. In addition, the layered troubleshooting approach allows easy and quick identification of the type of error or problem.

The layered troubleshooting approach provides various advantages as compared to general troubleshooting methods. The main advantages of the layered troubleshooting approach are as follows:

- Sequential analysis and identification of the problem
- Easy identification of the possible problem areas because the function of each layer is predetermined
- Facilitates improvement of network performance

Using the layered approach, you can troubleshoot problems pertaining to each layer separately. Let us understand different problems that occur on each layer of the OSI model and their possible solutions.

Troubleshooting Problems of the Physical Layer

The Physical layer of the OSI model is involved with transmission of a data stream in the form of bits. Data transmission between networks takes place through the Physical layer. This transmission takes place using various transmission media, such as:

Removable magnetic media: Provides maximum bandwidth for data transmission, and can be reused multiple times. Magnetic transmission media is prone to accidental destruction or destruction through natural disasters. Examples of magnetic media are floppy disks and tapes.

Twisted pair cables: Transmit data across long distances with the help of repeaters. An example of the most common application of twisted pair cables is the telephone system.

Coaxial cables: Provide high bandwidth and facilitate high-speed data transmission for longer distances as compared to the distances supported by twisted pair cables.

Fiber optics: Uses light signals to transmit data over long distances without using repeaters.

The problems of the Physical layer are related to the choice of the transmission media, which depends on the amount, distance, and rate of data transmission.

If data needs to be transmitted in small blocks over a short distance, and frequency of data transmission is low, twisted pair or coaxial cables can be used. However, if large blocks of data are to be transmitted over a long distance and with high frequency, optical fiber should be used.

In addition to this, data can be transmitted across networks using wireless media such as radio and microwave transmission. These transmission media are used to provide connectivity to mobile users. Unlike static transmission media, wireless media provides long distance communication and data transmission that spans the globe.

The Physical layer also deals with frequency and bandwidth problems of the data being transmitted. The Physical layer applies time-division multiplexing and frequency-division multiplexing to overcome problems related to efficiency of bandwidth utilization.

Troubleshooting Problems of the Data-link Layer

The main function of the Data-link layer is to ensure delivery of data from the Network layer of the source host to the Network layer of the destination host. While performing these functions, it is possible for the data to get corrupted or lost. To counter this problem, the Data-link layer uses the following protocols:

Unrestricted simplex protocol: This protocol assumes that the data is transmitted in only one direction, and that the Network layers of the source and destination networks are functional.

Simplex stop-and-wait protocol: This protocol assumes that the receiving Data-link layer stores data frames until these are transmitted to the Network layer. In addition, this is a simplex protocol, which assumes that the communication channel is error-free. The disadvantage of implementing this protocol is that it does not provide any mechanism to the recipient node to prevent overflow of data.

Sliding window protocol: Uses a single channel to transmit, control, and keep track of the data frames. It was introduced to provide bi-directional data communication and transmission.

The Data-link layer also uses various error-detection techniques, such as CRC and bit stuffing, to detect any type of transmission errors that corrupt the data frames. In addition, the Data-link layer applies a flow-control mechanism to prevent a fast sender from flooding a slow receiver with data frames.

Troubleshooting Problems of the Network Layer

The main function of the Network layer is to deliver data packets from the source to the destination network, using routers. The problems that can be encountered by the Network layer are:

- Overloading of a specific transmission route
- Inappropriate subnet topology
- Dependency of the Network layer on the subnet topology and its number
- Network congestion

To counter these problems, the subnet topology should be independent of the Network layer. Similarly, routing algorithms should ensure proper routing of the data packets from the source to the destination network. Examples of routing algorithms include flooding, shortest path routing, and flow-based routing.

In the *flooding* routing algorithm, each data packet is sent to every router. This results in the creation of multiple data packets. In the *shortest path* routing algorithm, a graph is created to identify the shortest path between the source and destination.

Both the flooding and the shortest path routing algorithms are based on the topology of subnets but do not take into account the load on each router. As a result, these algorithms do not provide accurate results. To ensure accurate results with respect to selecting the appropriate router, also consider the anticipated load on the router. Use a flow-based routing algorithm, which identifies the shortest route between the source and the destination networks, with minimum data transmission load.

Network congestion hampers smooth flow of data and may corrupt the data packets being transmitted. In addition, the Network layer applies congestion-control algorithms to counter problems of network congestion. These congestion-control algorithms monitor the network to detect the probable areas where network congestion can occur and apply corrective action to reduce the congestion.

Troubleshooting Problems of the Transport Layer

The Transport layer is the core of the OSI model because it serves as a bridge between the lower and the upper layers, thus ensuring reliable data transmission. The Transport layer can encounter the following problems:

- Unreliable connection between source and destination networks
- Unprotected transmission of data packets
- Delay in data transmission
- Error detection and correction
- Flooding a slow receiver with data packets

To overcome the aforementioned problems, the Transport layer uses the *three-way handshake* protocol. Using this protocol, the source host sends a connection request to the destination host. If the destination host is ready to receive data packets, it sends a connection acknowledgement signal to the source host. After receiving the connection acknowledgement signal from the destination host, the source host sends the data packet to the destination host.

On receipt of the data packet, the destination host sends an acknowledgement signal to the source host. If the transmitted data packet is lost during transmission, the destination host does not send an acknowledgement signal to the source host. In such a situation, the source host resends the data packet after a timer interval expires. Using this protocol, the Transport layer provides retransmission of the lost or corrupted data packets.

Troubleshooting Problems of the Application Layer

The Application layer generally encounters problems pertaining to data security during transmission. To protect data during transmission from unauthorized access and hacking, the Application layer applies various encryption techniques. Different encryption techniques include:

- Private key encryption
- Public key encryption

To implement *private key encryption*, the Network layer uses algorithms such as Data Encryption Standard (DES). In the DES algorithm, the data contained in the data packets is encrypted in blocks of 64 bits of cipher data, using a 56-bit key. The sender and the receiver agree on this 56-bit key before transmitting the data packets. After the data is transmitted, this 56-bit key is used by the receiver to decrypt the encrypted data.

To implement public key encryption, the Network layer uses the *Rivest-Shamir-Adleman* (RSA) algorithm. Using the RSA algorithm, the sender and the receiver use a pair of keys to encrypt and decrypt the data contained in the data packets. The key pair contains a public key and a private key. The sender encrypts the data using the public key of the key pair, and the receiver decrypts the data using the private key of the key pair.

Problem-solving Approach

Cisco has designed a problem-solving model that contains a list of troubleshooting processes and steps for resolving problems arising from a network failure. The problem-solving approach is a nine-step process, as listed here:

1. Problem definition
2. Fact collection

3. Possibility consideration
4. Action plan definition
5. Action plan implementation
6. Finding observation and review
7. Problem-solving cycle review
8. Error isolation
9. Problem resolution

1. Problem Definition

As a part of this step, ascertain the domain where the problem exists and determine the cause and scope of the problem. This will enable you to create an action plan in the subsequent steps of the Cisco Problem Resolution model. Figure 8.9 depicts a scenario in which intranets A and B are connected over a high-speed backbone. Computer 1 and Computer 2 are part of intranet A, and Computer 3 and Computer 4 are a part of intranet B.

In Figure 8.9, a problem arises when Computer 1 is unable to send data to Computer 4 due to disruption in the connection links between:

■ Computer 1 and intranet A
■ intranet A and intranet B
■ Computer 4 and intranet B

As a result, Computer 1 cannot establish a link with Computer 4. To resolve the problem, gather some facts about the problem. The output of the problem determination serves as an input to the next step, that is, fact collection.

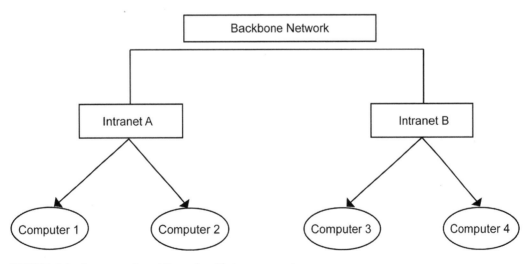

FIGURE 8.9 Intranets A and B, each with two computers.

2. Fact Collection

After you have determined the problem, the problem resolution model initiates a fact collection process. Gather all the information related to the problem, considering these factors and resources:

- Possible causes
- Event log entries from before and after the problem occurred
- Scope of the problem to estimate the probable effect on other areas of the network

Taking the example of intranets A and B, some basic questions, the answers thereto that enable an optimal process of resolving the problem are as follows:

- Is Computer 1 able to link to Computer 2?
- Is Computer 3 able to link to Computer 4?
- Is intranet A able to link to intranet B?

3. Possibility Consideration

After you have collected all the facts, analyze and select the one that is relevant to the problem. Using the example of intranets A and B, list different possibilities, such as the link between the intranets or links between hosts of an intranet are malfunctioning. This step helps to eliminate the possibilities not relevant to the problem in question.

4. Action Plan Definition

In this stage, document the steps to be taken to solve the problem. Select the most likely cause for the problem and create an action plan. An action plan contains a systematic procedure that lists the steps for resolving the error that has occurred. This step entails all the input of the fact collection step and consideration of possibilities. For example, consider the example of intranets A and B. The probability of a breakdown of the link between the two intranets is high.

When creating the action plan, perform an appropriate risk analysis. The action plan should not be set up in such a way that its implementation leads to other cascading problems.

NOTE

5. Action Plan Implementation

Action plan implementation is the step-by-step process of rectifying the problem by making changes. These changes are implemented such that any repercussions to a particular change can be easily traced. In addition, the actions performed for resolving the problem should be reversible in case of any subsequent problem. You also need to make incremental backups of data and configurations and maintain a log of the entire implementation process.

6. Finding Observation and Review

In this step, the final solution implemented for the problem is observed and reviewed. Analyze the changes made during implementation, any additional impact on the system due to the implementation of the action plan, and the alternate solutions considered during the action plan step.

This step is a review process for the previous steps of the model. The fact collection at the second step should be reviewed. Ensure that the information gathered for solving the problem is optimally utilized.

7. Problem-solving Cycle Review

The process of reviewing all the steps implemented and observing the results is called a *problem-solving cycle review*. The objective of this step is to distinctly identify the actual causes for the problem. In addition, you can enhance the problem-definition model and eliminate irrelevant changes.

8. Error Isolation

This step isolates the errors occurring within a system. Based on the input from the previous steps, isolate the actual problem or error from the system. Also, ensure that the entire process is logged, and that the error is time stamped to serve as a reference, in case the problem recurs.

9. Problem Resolution

As a part of this step, the identified problem is resolved using the solutions identified in the previous steps. After the problem is resolved, perform a test to confirm that the identified problem is rectified and the network is working properly. Then, document the entire process of resolving the problem for future reference.

CASE STUDY

David has recently joined Blue Moon Computers as an assistant network administrator. This organization deals with manufacturing hardware devices such as monitors, printers, and scanners. The head office of Blue Moon Computers is in Denver, with branches in Chicago and New York. Each branch office has an individual network that is connected to the head office. All of these networks are maintained and administered at the head office.

James, an employee of the Chicago branch, contacts David and tells him that he is unable to transfer a file to the New York branch. David consults his supervisor, Norman, who suggests that David should collect all the relevant information from James, identify the possible causes, and resolve the problem.

When David was exploring the causes of the problem, a probable cause could have been that a connection was not established between Host B of the Chicago network and the Host S of the New York network. After determining the problem, David should:

1. Collect facts and details about the possible causes of the problem. One cause can be overloading of the router causing the router to cease taking requests to transfer the data. Another possibility is that a connection was being established, but the data being transmitted was corrupted during transmission.
2. Identify the most likely cause of the problem. In this case, the most likely cause of the problem is that the router was overloaded and was unable to take any more requests to establish a connection and transfer the data.
3. Design an action plan to provide a solution to the cause of the problem. A probable solution of the problem could be for a priority algorithm being applied to all the requests being sent to the router. This would enable the router to prioritize the routing of data.
4. Implement a priority algorithm, such as a round-robin algorithm, to prioritize the requests.
5. Review the results of implementing the action plan.
6. Identify if there are any new problems that have occurred due to the implementation of the priority algorithm.

If reviewing the results does not indicate any other errors or problems, the problem is isolated, and its solution process is documented for future reference.

SUMMARY

In this chapter, we learned about network management systems and architecture. We also reminded you about network management industry standards and troubleshooting. In the next chapter, we move on to building a prototype and simulation plan.

POINTS TO REMEMBER

- The elements of the network management architecture include managed devices, management agents, management information base, network management applications, network management systems, and management protocols.

■ For fault management, administrators can employ event collectors such as Syslog servers, and event producers such as SNMP and RMON, in the network.

■ Performance management helps in mapping and maintaining performance of the entire network with the help of performance variables such as network throughput, user response time, and line utilization.

■ Network management industry standards include SNMP, MIB, and RMON.

■ The four basic SNMP commands are READ, WRITE, TRAP, and TRAVERSAL OPERATIONS.

■ An SNMP packet has two parts, *message header* and *Protocol Data Units* (PDUs).

■ The fields of a message header are version number and community name.

■ The five fields of a PDU are PDU type, requested, error status, error index, and variable binding.

■ SNMP v2 is an enhanced version of SNMP v1 and has two additional commands, GET BULK REQUEST and INFORM REQUEST.

■ SNMP v3 has some additional security features: NO AUTH NO PRIV, AUTH NO PRIV, and AUTH PRIV.

■ RMON was specified as a part of the MIB in RFC 1757 and considered as an extension of the SNMP protocol.

■ The key features of RMON1 are network fault diagnosis, planning, and performance management.

■ RMON2 is capable of monitoring the Application layer data traffic but not equipped to provide support to high-speed LAN/WAN topologies.

■ Network baselining is used to assess network health and performance, and provides an efficient technique to troubleshoot network problems.

■ The three troubleshooting approaches are Cisco hierarchical, layered troubleshooting, and problem resolution.

■ The Physical layer applies time-division and frequency-division multiplexing to overcome problems related to optimal utilization of bandwidth of the transmission media.

■ Different encryption techniques include *private key* and *public key* encryption.

■ To implement private key encryption, the Network layer uses algorithms such as Data Encryption Standard (DES).

■ To implement public key encryption, the Network layer uses the Rivest-Shamir-Adleman (RSA) algorithm.

9 Building a Prototype and Pilot

A *prototype* is a simulation of a network to be installed on the client side after a demonstration network has been approved. It can be considered as a trial run of the entire network design. The main purpose behind building a prototype is to get the network design validated from the client before the actual implementation.

To build a prototype model, find out the customer's requirements and expectations. Then, create a budget plan and get it approved by the client side. After this, implement the proposed design.

After determining the requirements of the new network, a prototype test is performed. In this test, a simulation is created with the help of real components, also called online testing. You can also test with the help of simulation tools such as NETSYS, which is called offline testing. In addition, software such as sniffers or protocol analyzers are used to examine the network traffic.

In addition to prototype testing, there is pilot testing and preproduction testing. The cost of testing is compared to its degree of risk. For example, if it is a small network, testing is not necessary because there is much less risk involved. If it is a large network, however, proper testing is required because it involves a lot of risk.

A pilot test consists of checking the installation and configuration of all the components of the proposed network. A prototype implementation and pilot test is carried out before a project is started, whereas a preproduction test is carried out after the project is approved. In preproduction, the test is done in detail for one or two components before it is put online. Successful testing of these components will reflect the success of the entire design. Therefore, during preproduction, it is necessary that tests run on test servers for a limited numbers of users. If the test is successful, the test will be replicated from preproduction to production servers.

Creating a prototype is most useful in the case of large networks, and building a pilot is more suited to small network designs.

DOCUMENTING DESIGN OBJECTIVES

The main purpose behind building a prototype or a pilot is to get the network design validated from the client end before actual implementation. Before moving to the building and implementation phases of networks, it is necessary to consider client needs such as performance, security, capacity, and scalability.

For this, document the design to create a proper step-by-step plan. The steps involved are:

1. Analyzing design requirements
2. Budgeting
3. Creating a design solution

Design Requirements

Design requirements are categorized as:

- Performance requirements
- Scalability requirements
- Security requirements
- Availability requirements

Performance Requirements

When you measure performance of a network, you are comparing its behavior to that expected in its design. The thought for a new design occurs when the existing network suffers from performance issues that accompany an increase in the size of the network. As the network grows, addition of new applications consumes system memory and gives rise to network bottlenecks. Therefore, while redesigning, the new design must be planned to satisfy the customer's requirement for the next three to five years.

For example, consider a bank with more than 2000 users who access their accounts through Internet banking daily. At any point, there are 150 users connected to the network. The current network has a lot of performance bottlenecks because too many users have logged on at the same time and have complaints about unavailability of the server.

As a network design solution expert, propose a solution to the client about the current situation. At the same time, estimate the number of users who would access their accounts through Internet banking three years from now. This is to avoid recurrence of the problem every two to four years.

Scalability Requirements

The new network should be scalable, in the sense that if the client wants to incorporate a new product or technology with the existing network, this design should accommodate those technical changes. Based on the projection of future needs, plan for design.

For example, if the client wants to incorporate three Web servers, one router, two HA firewalls, and a honeypot after the implementation of the new network, this network must be scalable enough to accommodate these additional components. This should be easily implemented in the new network. Redesigning the network would increase costs. Another alternative could have been to spend more money every year for enhancing the current network or for designing solutions, which may not be always possible.

Security Requirements

Security is an important consideration while designing a network. With instances of network failure through intrusions and viruses on the rise, the network has to be protected against security threats. Check out whether all security components such as firewalls, Intrusion Detection Systems (IDS), honeypots, and virus-detection software are used adequately. Using such technologies would protect and immunize the network, but should not hamper the workflow. It is recommended that all your security products should be performing only their specific function. For example, do not employ too many security products from different vendors on the network unnecessarily because this can complicate the whole network. If we are using a firewall from one vendor for filtering viruses, there is no need to use an additional firewall from a different vendor for the same purpose.

The security considerations to be kept in mind while redesigning are as follows:

- Check the network connectivity with the external network.
- Check the gateway security and its effectiveness.
- Check the perimeter security.
- Check for user-level security and user awareness about security policies and procedures.

Figure 9.1 shows a scenario in which you have two Web servers, one proxy server, one mail server and a firewall gateway, all interconnected to protect the data within the internal network of an organization.

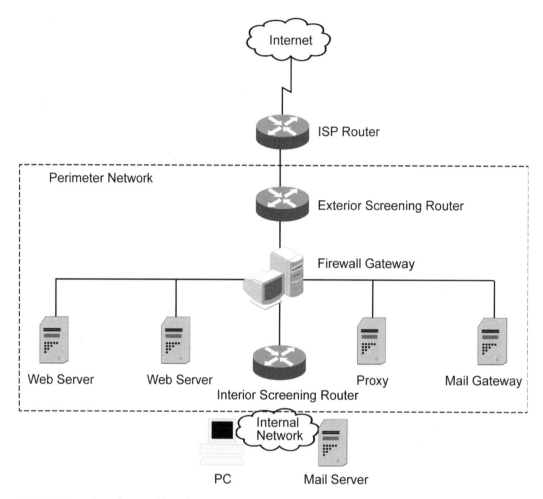

FIGURE 9.1   Security considerations.

Inserting an additional firewall between the interior screening router and internal network will generate unnecessary traffic, causing delays in data transfer. If your problem is resolved without adding an additional firewall then do not complicate it by adding unnecessary security devices.

Availability Requirements

Currently, with the increasing demand for 100 percent uptime of servers, it is essential that your network resources should be available around the clock. While designing a network, design a backup solution. In case of network failure, whether it is a single point failure or of the whole network, the traffic flow should not stop. To overcome the problem, create a disaster recovery mechanism for the network. For example, in a banking environment in which transactions are made 24 hours a day, it is important to keep the network up all the time. For this, you should have at least two backups for each technology or product.

For example, if the leased line of the Internet banking section fails, a router stops functioning, or all the links of the vendor are disconnected, your network design should be able to handle these discrepancies through backups. In addition, you can provide an ISDN connection as the primary backup for the leased line and satellite link (VSAT) as the secondary backup. This way, you can keep the situation under control without affecting the organization's normal flow of work. Another option can be a disaster recovery site solution. A disaster recovery site functions when your prime location is destroyed or is not working.

Budgeting

Budgeting is an important part of the design. After designing the network, calculate the cost of the design. Budgeting depends on the type of business and data flow. Try to keep the budget low with the maximum possible output.

Sum up the total cost involved by finding the product cost, license fee, equipment installation/mounting cost, and so on, and then check if you can provide the same solution at a lower cost.

For example, Cisco 1700 Series routers deliver a comprehensive suite of integrated security capabilities with wire speed IPSec VPN, stateful firewall protection, and intrusion detection. They also provide a migration path to Voice-over-IP (VoIP) and IP telephony services. Now, consider a small sized organization with a Web server that shows only static pages, and may not require high-speed leased lines or dedicated firewall, routers, or IDS. If your client is an ISP, then their setup may include a high-speed communication line, HA firewall, and high-end routers.

NOTE

Stateful firewalls maintain the status of all connections established through them.

While budgeting, the important points are:

- Count the current network assets.
- Determine if you can use them for the new design.
- Determine if you require an extension of the old design or a new design.
- Count the number of network assets you still need to purchase.
- Try to find a low-cost solution by buying items that can be used for multiple purposes.
- Ascertain if there are licensing fees for operating system software or other applications and databases.
- Calculate the total cost and see if it fits in the budget.
- Preferably, try to make purchases from a single vendor to take advantage of discounts on products and services.

Design Solution

After we have analyzed the client's requirements, the last phase is designing the solution. Evaluate all the data. If something is missing, consult the design consideration explored in the requirement phase for the missing steps.

To create a design solution, draw a design diagram and see if it is adequate. Determine if all the aspects of the design objectives have been carried out. If they have been, finalize the design for presentation to the client.

Here is a summary of the design procedure:

■ List details about the existing topology and then choose a new one, if necessary, that will best suit the client's needs.
■ List all the network and security components, that is, make a rough model/mapping of routers, switches, firewalls, IDS, Web servers, database servers, and application servers used in the design.
■ List all the standard IOS software and all operating systems that should be used.
■ Assign priorities as designated by the client. If the project is divided into phases, provide complete detail about all the phases, activities involved in each phase, and the cost involved.
■ Make a list of each task to be undertaken in each phase and the cost of each.
■ List your alternatives, that is, alternate products or technology if required. Sometimes, alternate technology can serve as a backup solution if the prime solution cannot be achieved. It is always advisable to come up with a backup solution.

Take an example of a bank that is undergoing technological changes to incorporate its Internet banking into the normal banking routine. This is done in three phases:

Phase 1: Collecting data about the existing network infrastructure for the client. For example, explore the client's existing network by finding answers for the following questions:

■ Does the bank already have a WAN setup?
■ If yes, then what are the network components used, and why are they used?
■ What is the current number of bank customers?
■ How much do they expect their customer base to increase in the next five years if Internet banking is started?
■ How much time is alloted for building the new network for Internet banking?
■ How much money has the bank approved for its new setup?

- What criteria (such as availability, security, scalability, and performance) are required to complete a full planned project?

After finding these answers, see if you can use some of the old technology or products. If not, then plan for a new design. Check out the products and technology that would be relevant to the client's needs and requirements.

Opt for a single router placed at the point of the intranet connecting it with the Internet. The outer router (IB-1) will be connected to a switch (SW-1). This switch should be such that it can support multivendor routers. A firewall (F-1) and IDS (IDS-1) will also be connected to this switch.

The firewall is connected to two more switches, one switch (SW-2) which will be used by Web servers and the SSL server. The other switch (SW-3) will connect to the application and database servers.

Figure 9.2 shows the first phase of the design process.

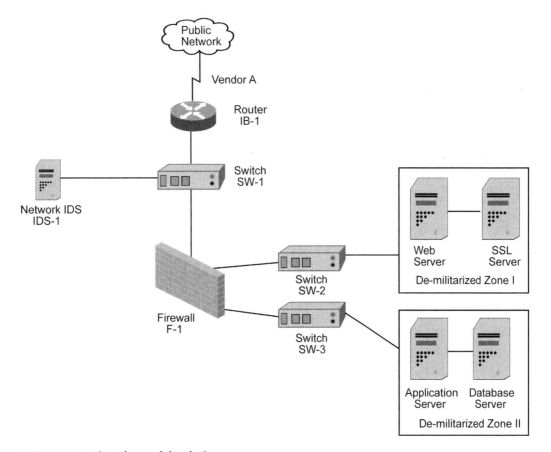

FIGURE 9.2 First phase of the design process.

Phase 2: This phase involves some backup measures to retrieve the organization's data in the event of some failure.

For example, if router (IB-1) fails, then one more router (IB-2), which will be from a different vendor, should handle the traffic flow. For this reason, in the first phase, we had a switch to support an additional router from a different vendor. This was responsible for assigning a virtual IP address that is shared between the routers acting as a redundancy pair. Also, in the second phase, use of an HA firewall (HA-F-1) is planned to help load balance. In case, if one firewall stops working, the other would take over. In addition, plan for extra cluster servers for the Web server or the database server.

Figure 9.3 shows a complete network design including all phases involved in designing networks.

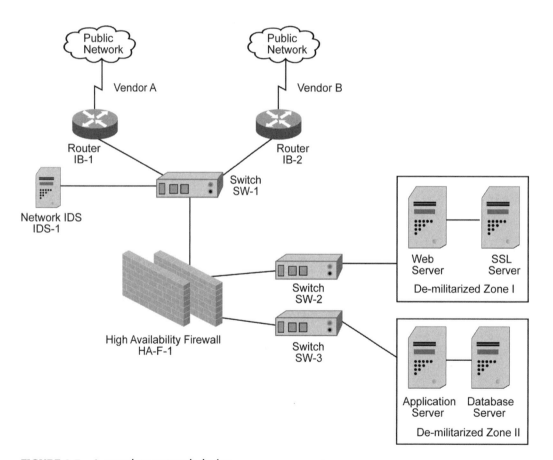

FIGURE 9.3   A complete network design.

Phase 3: The last phase covers safety measures required for a business continuity site and disaster recovery site. This is useful in situations such as a database crash. For such a situation, disaster recovery software would be helpful in maintaining the network continuity.

After discussing the design consideration as per the client perspective, build a prototype to demonstrate how the client's needs have been covered.

PROTOTYPE MODEL AND PILOT

The difference between a prototype and pilot is that pilot demonstrates only some of the main functions of the whole network as per customer needs, whereas a prototype demonstrates complete functioning of the designed network as per the customer's needs. Due to this, a prototype needs more planning and investment in terms of procuring the equipment and other infrastructure to demonstrate, whereas pilot can be demonstrated within the help of minimal equipment and infrastructure.

Table 9.1 lists the major differences between a prototype and a pilot.

TABLE 9.1   Differences Between Prototype Model and Pilot

Prototype	Pilot
Used in large networks, designed on a subset of a bigger network.	Used in smaller networks such as simple WAN networks or one or more network segments.
Used to simulate complex functions such as data transfer between connected systems.	Used to simulate basic functions such as connectivity.
Cost is higher as it is designed for large networks.	Cost is lower as it is designed for smaller networks.
This is the choice when a client requires a full run of the network before full implementation	This is the choice when the customer needs a minor test of the network design.

Building a Prototype Model

To plan a prototype model, there is a procedure to be followed. The sections relating to the customer network requirements and information act as an important outline for building a prototype. You can create a final prototype based on these guidelines:

1. Assessing customer needs
2. Setting prototype model requirements
3. Determining the size of the prototype
4. Studying about major competitors
5. Developing a test plan
6. Procuring equipment
7. Demonstrating the prototype model

Assessing Customer Needs

Assessing or analyzing customer needs is important for understanding what the customer requires in terms of technology, reliability, and usefulness. Interviewing the management and users about their infrastructure needs helps in acquiring this information. This includes a set of questions on the project time lines, resource availability and requirements for a prototype, and budget limitations.

Setting Prototype Model Requirements

Documentation done in this phase includes technology requirements of a prototype model including any additional requirements in terms of equipment or communication links. A network designer at this stage has to decide how the present infrastructure can be used to design the new network, keeping in mind the new requirements. The documentation can help to streamline the budgeting requirements and other complexities in design. To make the documentation easier, use a sample report in the form, as shown in Table 9.2.

TABLE 9.2   Prototype Requirement Chart

	Infrastructure on Location 1		Infrastructure on Location 2	
	Present	**Additional**	**Present**	**Additional**
Routers				
Switches				
Security devices				
Other network devices				

Determining Size of the Prototype

The main purpose of determining the size of the prototype is to decide the extent of replication of the original network to prove that the original network design goals are met. For this, determine:

The current and future health of the network: If the future strength of the network is expected to be increased by, say 500 users, then the prototype model should include at least 50–100 users.

The future security expectations from the customer: If the customer decides that his network should be further secured against viruses or hacks, provide the client with additional firewalls and IDS in the prototype mode.

The approximate cost for the whole project: Throughout the designing phase of the prototype, budget will be a consideration because the new network design would be based on this model. Care should be taken to make sure the project doesn't go over-budget.

For example, if the original network has an Oracle product as the central database while working with 4 servers, 1500 users, and 20 switches, you can make the following changes: install one server with the Oracle product and have 2 switches with 50–100 users for the prototype model.

Studying About Major Competitors

A customer planning to opt for a new design and infrastructure is not restricted to a single vendor. It is important to know the competitors and analyze their design and infrastructure because the product used in the design can vary for each competitor. This would give an edge over other current competitors. For better presentation of the prototype, generate a report to provide a broad view of comparison with products from other competitors.

Developing a Test Plan

In this phase, two different plans are developed. The first plan presents the technology in terms of the new design, topology map, changes in addressing scheme, and changes in any application flow. The other plan presents the resources in terms of who owns the project, who would be accountable, and resources such as design and implementation engineers. This phase also presents a Gantt chart, which shows total resources, total time required for project, and when and where the resources are utilized in terms of man-hours so that the client gets a complete projection of the entire project.

Now, present the documentation of this phase after crosschecking technology resources and determining that all the requirements of the customer are met. The high-budget resources such as communication links or core devices have to be checked for optimal utilization before presenting in the test plan.

Devise a test plan to:

- Describe and demonstrate your activities and the tests that will be executed in the prototype test.
- Give the customer a feel of the prototype.

In addition, it is important that you practice your test plan before showing it to the customer.

Procuring Equipment

In this phase, after finalizing the test plan, the equipment is procured and configured for building and implementing the prototype, and it is then prepared for the final phase.

The purchases can include the following hardware and software devices:

- Simulation tools
- Cisco hardware and software devices
- Hardware and software devices from different vendors

Demonstrating the Prototype Model

The final phase is when the prototype model is demonstrated to the customer. Maintain a checklist of the customer needs and obtain the feedback.

Building a Pilot

Building a pilot is different from building a prototype because it addresses the needs of smaller businesses or networks as compared to a prototype. Building a pilot includes some of the phases that are covered in a prototype, but differs in the way the needs are demonstrated. Some of the guidelines/phases for building a pilot are:

1. Testing the design: This means that a pilot should be tested at the designer's premises before being test-run at the client side.
2. Investigating client proposals: This step is performed similarly to the *Assessing customer needs* step of prototype building.
3. Designing a script for the pilot demonstration: The script encompasses details of the test results. Some points highlighted in the script are:

   - Verifying if the results of the conducted tests satisfy the client's specifications.
   - Reviewing the success of the equipment used in the pilot demonstration.
   - Demonstrating where the competition fails.

4. Practicing: Involves preparation for the final presentation at the client end ensuring that things work out as planned.
5. Presenting the pilot: This step is same as the *demonstrating the prototype model* step of prototype building.

Before submitting the prototype/pilot on the client side for approval of the final network design, the best practice would be to test both at your end. This step is necessary to judge that the prototype/pilot is up to the standards as per the specifications.

Cisco manufactures certain tools for the purpose of testing the prototype/pilot. These are:

- Cisco IOS commands
- Protocol analyzer

Cisco IOS Commands

These are commands used to test the functioning of the designed network. It has a set of commands for testing the network infrastructure.

The Show Command

The show command can be executed in the user and privilege mode. Cisco IOS provides a range of show commands that display information about the rate of utilization of router resources, network interface status, and router configurations. You can isolate problems and determine the exact cause of performance slowdown or failure using show commands. The different show commands are as follows:

Show version: Provides the IOS version and its internal name. Using the internal name, you can display the hardware configuration, such as processor type, memory size, and existing controllers.

Show startup config: Displays the router configuration stored in the NVRAM (Non-Volatile RAM). This information is useful if the router configuration is changed during the session. You can determine the configuration of the router during bootup.

Show running config: Displays the currently active configuration for the router. The show running config command is used to isolate problems with the router or reasons for a crash.

Show interfaces: Displays the status of all the interfaces configured for a router or access server. The output of the show interfaces command depends on the version and type of router being used.

Show controller: Displays the network interface card controller statistics. Depending on the type of interface of the network, the output displays various details including the microcode. *Microcode* is a set of instructions or code written in low-level language, used for a NIC.

Show flash: Displays the content and layout of the flash memory, which includes information about the IOS software engine.

Show buffers: Shows buffers in the routers that are allocated from the shared system memory to store packets during process switching. At times, some buffer settings require tuning to synchronize process switching and maintain a standard system performance.

Show memory: Displays memory pool statistics and information about the activities of the system memory allocator. It displays a block-by-block list of the rate of memory use. This command is useful when router performance is a problem area.

Show process CPU: Displays active processes on the router along with the corresponding process ID and status of the priority scheduler test. It displays the CPU utilization of the router, the CPU time used by the router processes, and the number of times the CPU was invoked.

Show stacks: Displays the status of stack utilization of processes and interrupt routines. It also displays the reasons for the last system reboot. It is used when a system fails, because it displays information about the failure type, failure program counter, address, and the stack trace of the operand, which is stored by the ROM monitor.

Show CDP neighbors: Displays information about the neighboring devices directly connected to the router. It provides *reachability* information, which helps determine the status of the devices at the Physical and Data-link layers.

Show debugging: Determines the type of debugging enabled on a particular router. This information is useful for switching to a different type of debugging mode or to disable a particular type of debugging when more than one debug module is running.

Show logging: Displays the status of the syslog errors and event logging. It displays the host addresses, the type of logging being performed, and various logging statistics including the messages stored in the log buffer.

Process switching is the method in routers where every packet is processed in the router software.

NOTE

The ping Command

The ping command checks the connectivity between nodes on a network. It sends ICMP echo messages to the destination node and waits for a reply. If a reply to the echo message is not received, it confirms connectivity problems on the network. The ping command can be executed in two modes, the *user* mode and the *privileged exec* mode. In the user mode, default parameters are set to check for connectivity. For example, five echo messages for 100 bytes each are sent by default to the destination host with a timeout interval of 2 seconds. While troubleshooting connectivity problems, it is important to identify the problem area. If the ping command does not receive a reply from the destination host, it should be used for destination hosts that are nearer to the source host. As soon as a reply is received from a particular destination host, isolate the exact problem area on the network. Connectivity problems can occur when the devices are extremely busy or have restricted access on the network. In case of a connection timeout or devices with restricted access, identify the busy or prohibited devices and perform remedial action accordingly.

The debug Commands

Debug is the basis for a series of advanced commands that provide high-end options for monitoring and retrieving network data for troubleshooting. They can be executed only in the privilege exec mode. The debug commands are used only for temporary or specific troubleshooting purposes because they are highly resource-intensive. The debug commands prevent high-speed switching of data packets and force the use of a route processor. These commands process switching before data packets can be sent to the interfaces from where they are finally dispatched. This reduces the speed of operation of the router and increases the processing time, which in turn, reduces network performance.

However, during troubleshooting, the debug commands are used to collect data for each data packet that is transmitted. As a result, it forces all the data packets to be process switched, which consumes a great deal of router resources. As a result, the debug command should be used with caution while troubleshooting. To isolate problems and derive alternate solutions, you can use the debug cdp packets command to debug the packets transmitted from one host to another using the current router.

Table 9.3 lists the CISCO IOS commands and their description.

TABLE 9.3 CISCO IOS Commands

Command	Description
Show Interface	Helps display Data-link layer errors, routers errors, and broadcast rates
Show Process	Displays router CPU use and time taken to complete the processes
Show Buffer	Displays the status of the buffer
Ping and Traceroute	Help in troubleshooting connectivity and performance issues
Show Protocol Route	Helps in listing the protocol routing table for troubleshooting routing problems
Show Access Lists	Helps in displaying access lists for troubleshooting security-related problems
Debug	Helps in troubleshooting and verifying sent and received packets

Protocol Analyzer

Protocol analyzers are software programs that function by recording, interpreting, and analyzing operations of a protocol within a network. In addition, they filter traffic from a particular device and generate frames for transmission over the network. Protocol analyzers do not affect network performance. They provide details about network traffic and protocol paths, network configuration and operation, and offer potential solutions to critical network problems. Some examples of protocol analyzers are:

■ Network Associates' McAfee® Sniffer® suite of products
■ Radcom's product series of protocol analyzers for LANs, WANs, and ATMs

When a node transmits frames on the network, the protocol analyzer captures the frames and decodes protocol layers in a recorded frame. This information is in the form of a readable summary, which provides information about the protocol layer along with the function of each byte in the frame. As the network size grows, the scope of the functions performed by the protocol analyzer increases because they need to detect and decode all the protocols used by the frames.

The protocol analyzer also generates and transmits frames for capacity planning and performing load tests on the network. For example, if network performance repetitively deteriorates at a particular time or in a particular region, one of the reasons can be heavy network traffic during that period or in a particular region. To detect and reduce such performance-related issues, the protocol analyzer should be able to send multiple captured frames.

A protocol analyzer works in two modes, capture and display. In the capture mode, it records frames of the network traffic depending on certain performance criteria or a preset threshold. For example, you may observe network downtime when data is transmitted to a particular network. In order to determine the exact cause of the network failure, attach a threshold or filter to the protocol analyzer to capture the frames directed to that particular network. In such a situation, the protocol analyzer captures the frames directed to that particular network in the capture mode. The captured frames will have a timestamp attached to them and will determine the exact period during which the network performance deteriorates. This type of information is critical for organizations that require seamless network connectivity, such as banks and stock exchanges.

In the display mode, a protocol analyzer decodes the captured frames and stores the information in a readable format for future interpretation. To view the captured frames, you can use thresholds. In addition, apply these thresholds to view only those frames that match a certain criteria.

Protocol analyzers are intelligent tools that use specialized techniques to diagnose problems based on the symptoms. The knowledge bases of protocol analyzers include the following:

Theoretical databases: Store information about the standards

Network-specific databases: Store information about the network topology

User experience: Includes records of the network problems that have occurred within the network

SUMMARY

In this chapter we learned how to build a prototype and pilot by documenting our design objective and understanding the design requirements. We also reminded you about guidelines for a good prototype and verifying design objectives.

POINTS TO REMEMBER

- A prototype involves a simulation, which is created with the help of real components. Prototype testing is also called online testing.
- A pilot test consists of checking the installation and configuration of all the components of a proposed network.
- Creating a prototype is useful in large networks, and building a pilot is useful in small network designs.
- The steps involved to document design objectives are analyzing design objectives, budgeting, and creating a solution.
- Design requirements include performance, scalability, security, and availability.
- A disaster recovery site functions when your prime location is destroyed or is not working.
- A pilot demonstrates only some of the main functions of the whole network as per customer needs while a prototype demonstrates complete functioning of the designed network as per the customer needs.
- A prototype needs more planning and investment in terms of procuring the equipment and other infrastructure as compared to a pilot.
- The guidelines for building a prototype include assessing customer needs, setting prototype model requirements, determining the optimal size of the prototype, studying major competitors, developing a test plan, procuring equipment, and demonstrating the prototype model.

- The guidelines for building a pilot include testing the design, investigating client proposals, designing a script for the pilot demonstration, and practicing and presenting the pilot.
- To test a prototype/pilot, use Cisco IOS (commands) and protocol analyzers.
- Protocol analyzers filter traffic from a particular device and generate frames for transmission over the network.
- A protocol analyzer works in two modes, *capture* and *display*.
- The knowledge base of a protocol analyzer system includes theoretical databases, network-specific databases, and user experience.

Appendix ▓ **About the CD-ROM**

The CD-ROM included with *Designing Networks with Cisco* includes all the code listings, tables, and images from the various examples found in the book. In addition, this CD-ROM includes an exhaustive Question Bank for the readers to test knowledge acquired after reading the book.

CD-ROM FOLDERS

Listings: Contains all the code listings from examples covered in each chapter.
Figures: Contains all the figures from each chapter.
Tables: Contains all the tables from each chapter.
Question Bank: Contains multiple-choice questions to test the reader's knowledge on Cisco designing concepts.

OVERALL SYSTEM REQUIREMENTS: RECOMMENDED

- Windows® 98, Windows NT, Windows 2000, or Windows XP Professional
- Pentium III processor or greater
- CD-ROM drive
- Hard drive
- 128 MB of RAM, 256 MB recommended
- 20 MB of hard drive space for the code examples

INSTRUCTIONS FOR USE OF THE QUESTIONBANK

1. Insert the CD-ROM in the CD-ROM drive
2. Double-click the QuestionBank folder
3. Double-click the Designing-Setup.zip file

4. Extract the setup files in a folder (example, create a folder called DesigningQB in your C, D, or any other hard disk drive)
5. Double-click the DesigningQB folder
6. Double-click the Designing-Setup folder
7. Double-click the setup.exe file
8. Follow the setup instructions
9. Once the setup is over, you can access the Question Bank from the Start→ Programs→Designing Networks with Cisco

Index